Material Vernaculars

MATERIAL
VERNACULARS

Jason Baird Jackson, *editor*

Material Vernaculars

Objects, Images, and Their Social Worlds

Edited by Jason Baird Jackson

Indiana University Press, in cooperation with the

Mathers Museum of World Cultures, Indiana University

Bloomington and Indianapolis

This book is a publication of

Indiana University Press
Office of Scholarly Publishing
Herman B Wells Library 350
1320 East 10th Street
Bloomington, Indiana 47405 USA

iupress.indiana.edu

© 2016 by Jason Baird Jackson

A free digital edition of this book is available at IUScholarWorks: http://hdl.handle.net/2022/20925.

The paper used in this publication meets the minimum requirements of the American National Standard for Information Sciences—Permanence of Paper for Printed Library Materials, ANSI Z39.48-1992.

Manufactured in the United States of America

DOI: 10.2979/materialvernaculars.0.0.00

Cataloging information is available from the Library of Congress.

ISBN 978-0-253-02293-6 (cloth)
ISBN 978-0-253-02348-3 (pbk.)
ISBN 978-0-253-02361-2 (e-bk.)

1 2 3 4 5 21 20 19 18 17 16

Contents

Material Vernaculars

Jason Baird Jackson

Material Vernaculars: An Introduction

A BOOK SERIES about which I am very hopeful and a collection of scholarly essays that I hope you will engage and profit from—introducing them both is my task. They are twin hopes twisted like two fibers wound around each other to produce what I intend as a stronger piece of string. Extending the metaphor, I can observe that the techniques by which the world's peoples produced cordage were among the earliest and most fundamental matters that the then young fields of anthropology and folklore studies concerned themselves with. We often speak of string as a "simple" thing, but early work in my fields understood its production by varied means and of varied materials as a vital accomplishment and as a widespread human necessity. Around the world, string was—and is—the basis for cultural elaborations of a near infinite sort. Think about fishing nets, home building, bird cages, weaving, bridge-building . . . the list goes on and on, but think also, for instance, of string figures—a classic ethnographic topic if there ever was one—which can underpin such arts as storytelling and can contribute to such crucial dynamics as the socialization, and the amusement, of children. While my own children are more likely today to be amused by a game or story presented on a mobile phone, they—middle-class children from the middle of the United States—can still make a cat's cradle, and they learned how to do so informally from peers. That this is so points to one intended purpose for this series devoted to understanding material and visual culture within particular social worlds. Older concerns in the older fields of material culture studies have not stopped being relevant in a world of mobile phones, server farms, and genetically modified salmon. While I have no particular author in mind, the Material Vernaculars series will, I hope, be the perfect series in which a careful, thoughtful, sophisticated study of handmade (perhaps soon, we might say "artisanal") string might be

published. More generally, the series is intended to extend and re-
new traditions of material culture study that reach back to the beginnings of the field—particularly as manifest in the disciplines of
ethnology, cultural anthropology, and folklore studies.

If the series is to be a good home for long-form scholarship on, for
instance, local weaving markets or regional vernacular building practices or the making and uses of kites, the series editorial board and
I intend though for the opposite to also be true. While there are
societies today where locally produced string is made according to
local techniques—and the same is true of so many other things—
string bags, hoop nets, hammocks, woolen blankets, cotton fabric,
and so on—we, the people of the twenty-first century, do live in
changed material circumstances from those encountered by the ethnographers of the nineteenth and twentieth centuries. The way that
my children can huddle in culturally specific ways with their friends
around a mobile phone to play a video game according to informally
transmitted norms and in light of local social dynamics is a token of
the fact that material vernaculars are by their nature ever changing
and thus continuously open to new kinds of investigation and interpretation. Much of the most exciting work in the fields of material
culture studies over the past two decades has sought to approach
innovative and, in some measure, new material culture values, forms,
and practices through innovative research strategies, such as through
multi-sited fieldwork, through the study of commodity chains, and
through attention to new media, new manufacturing techniques, and
new consumption habits. A full accounting of trends and approaches
in contemporary material culture studies is beyond the scope of this
introduction (Berger 2009; Geismar 2011; Glassie 1999; Löfgren
2012; Miller 2010; Shukla 2008, 383–429; Tilley et al. 2006). It is
possible to note though that there is something of a disconnect
between scholarship—centered in museums, for instance—that
builds on the older scholarly traditions of studying what are now
older material forms and their resonances and newer research lines
that do not stress these continuities but instead endeavor to find new
approaches to the study of new phenomena. This distinction is drawn
too starkly, and admirable workers who defy this binary can readily be
found—consider, for instance, Smithsonian curator Joshua Bell's
work on my two iconic examples: string figures (Bell 2010) and mobile phones (Kemble et al. 2015). But I trust that the distinction and

the isolation of these scholarly worlds will be recognizable to my colleagues in ethnology, cultural anthropology, and folklore studies.

It is my hope that the Material Vernaculars series will not only serve as a robust publishing venue for work on either end of this continuum but that it will successfully cultivate work that brings them together, expanding the space of overlap in a Venn diagram that I have tried to draw here in words. It is my aspiration that the spirit of the old and the new, and their integration, is captured in the cover image selected for this volume. That photograph, of a basket seller encountered in a small city in a rural district of Guizhou province in Southwest China, brings together handmade bamboo baskets intended for use in everyday labor (a classic "old" topic) with the ubiquitous multifunction mobile phone (a classic "new" topic). Some of the baskets shown were rerouted—through my collecting—into the holdings of the Mathers Museum of World Cultures (setting up their study as a classic "old" topic), but this encounter took place in the context of a large, multinational, multi-museum project aimed at grappling with the new roles played by museums in local, regional, national, and international "heritage regimes" (a classic "new" topic).

I offer these words at a beginning and I am cognizant that series editors, like the authors of ambitious manifestos of all sorts, can be left looking foolish when history shows that grand hopes have gone unrealized. As the proverb goes (or used to go) "the proof of the pudding is in the eating." I will come to this collection of essays in a moment but can point happily to the first full-length monograph to be published in this series. As I write these words, that book is in production and is slated to appear alongside this collection. It is Jon Kay's study *Folk Art and Aging: Life-Story Objects and Their Makers*. You can get a taste of this fine book through Kay's essay, included in this volume, but I hope that you will read the full study for yourself. It is a beautifully crafted piece of research that compellingly conveys the rich role that objects, and narratives about them, play in the lives of contemporary American seniors. It will, I believe, be greatly valued by Americanist ethnographers and by those interested in vernacular art traditions, but the study offers much to an audience with little exposure to the study of material culture—gerontologists and others who develop support and services for seniors in the United States and other aging societies. Kay illustrates compellingly how in-depth attention to the creative lives of seniors can inform better social policy and caregiving practices. For material culture studies scholars, the book also illuminates the important ways that the making of objects are bound up with the narrative lives of their makers. I will return to this theme below.

Beginnings pose special risks. As tangible works, *Folk Art and Aging* and the collection of essays gathered here may unduly shape readers' and potential authors' early sense of the series' intended scope. Wanting to keep our vision open and plural, I have tried to write against this urge here. As I noted at the outset, this introduction addresses not only the series but the specific essays that have drawn your attention to this volume. While in no fashion bounding the scope of the series that they help inaugurate, the essays gathered in this collection do fit together and complement each other in ways that I would like to highlight. While many issues of great concern to our fields and to me as this series' editor are absent from these essays—the themes that they do take up and the connections that are readily drawn between them are just the kinds of issues that I hope that potential series authors will feel encouraged to address in their own future work.

What are these shared themes? One concerns the ways that images and other objects of material culture are actively produced and performatively used—recursively and creatively—by individuals concurrently in support of oral communication in verbal genres such as personal experience narrative or sacred narrative. In varying degrees, this theme is present in all of the chapters gathered here. It is particularly overt in Kay's chapter on the objects made by Indiana seniors and used in tandem with life-history narratives. Images and objects similarly bring to life the family stories narrated (or elided) by the makers of elaborate scrapbooks. As discussed by Danille Christensen, the scrapbooks of the contemporary United States share this quality with many of the ledger drawings at the center of Michael Paul Jordan's account of women's involvement in warfare among the Kiowa people of the nineteenth-century American Great Plains. As a rich body of literature on such drawings has discussed, they represented a transformation of older modes of men's representational art on the Plains. While such art was once publically viewable on surfaces such as tipi (tent) covers, clothing, and rock art, the (then) new technology of the bound paper ledger provided surfaces on which personal experiences could be pictured and later brought out and narrated to an intimate audience. Such practices share much with the instances described by Kay and Christensen. The theme of material form connecting with verbal exegesis also runs through Gabrielle Berlinger's study of the meaning of Sukkot for Jews in a range of social settings. In the case of continuity and change in the wedding clothing of the Osage examined by Daniel C. Swan and Jim Cooley, oral history is one part of the expressive cultural context that encompasses and informs local uses of these special clothing objects, but the authors attend to a wider range of concurrent phenomena, including marriage practices, economic relations, ceremonialism, and music and dance performance. Especially in folklore studies but also in an increasingly specialized field of cultural anthropology, students of material culture have not tended to be, for instance, also scholars of verbal art or of ritual. These essays affirm the value of looking at the intersection of such realms, particularly at the ways that people use objects alongside other communicative modes.

Issues of public and private run through the essays as well. While older types of arranged marriage among the Osage were negotiated by families in private, they were validated in very dramatic and very public ceremonies. Fine, beautiful clothing of a very Osage sort was fundamental to both the private and public sides of marriage customs, and these distinctions carry forward into the adapted contexts in which such clothing remains vitally important to Osage people. Privacy and publicity similarly are present in all of the studies gathered here. Sukkot practices manifest high variability in part because the religious holiday operates at a family or household level rather than being a more centralized congregational rite. Individuals and families have unique opportunities to cultivate and perpetuate their own idiosyncratic practices and interpretations while also connecting—through hosting guests, for instance—to shared local and extralocal vernaculars. The public-private dynamic was already noted in the case of ledger book drawings such as those executed at Fort Marion by Kiowa artists, but it holds true for the case of scrapbooks and life-history objects too. As Kay notes in his study, making such tangible objects can be an important kind of solitary memory work, but it is a memory work for which, as with scrapbooks intended as gifts for loved ones or as objects to be explicated to an audience verbally, a more social (if still small-scale) use is anticipated.

Movement and variation across time and space—long a concern in material culture studies—are prominent here, although they are explored at different scales and with different emphases by all of the authors. Two works are overtly historical in the historian's most familiar sense. Swan and Cooley's study of Osage wedding clothing—particularly the emblematic wedding coat derived and localized from a European military prototype—tracks the introduction of this garment in the moment of colonial encounter but then illuminates continuities and changes within the specifics of Osage community life up to the present day. Similarly, working in the realm of historical ethnography, Jordan aims to illuminate the important domain of women's values and roles in a specific period among the Kiowa. An end unto itself, this historical query also challenges assumptions woven into a larger body of historical literature on Native American gender roles. At the same time, Jordan is also examining practices by which Kiowa artists of the past recorded their own history. While primarily ethnographic, Berlinger's study of Sukkot is at its core bound up with similar questions of historical consciousness, that is, in her case, how contemporary Jews understand and reanimate a past that, for them, stretches back to ancient times—times that are, nonetheless, very much alive and relevant in the present. In the quick time of contemporary life, Christensen's very recent ethnography nonetheless demands historical calibration, as when she nods to the way that scrapbook work is not quite as vital in 2016 as it was just a few years ago. Christensen herself is, in that frame, offering a kind of history of the near past. More obviously, her study is about the way that scrapbook makers —mainly women—record, narrate, and perform their own histories for personal, family, and, sometimes, public purposes. As you read the chapters, I urge you to keep this range of historical concerns in mind.

Similarly, readers will profit I think from considering the physical movements of the people, objects, practices, and values that the authors discuss. The movement of Sukkot architecture, ritual, and practice across space as well as time is extraordinary, illustrating its flexible and compelling nature. Similarly, the Osage wedding coats represent a breathtaking value transformation as they move from being objects of European male material culture of the past (associated with places such as Washington or London) to Osage female material culture of the present (associated with rural Osage County,

Oklahoma). Broadly speaking, the material forms of concern to Kay and to the elders with whom he has studied are not unique to them, to their communities, or to their home state of Indiana. As forms at least, they come from other places and have careers in time and space. Yet, as with the Osage coats, they have been localized and in a sense vernacularized. The same can be said of the scrapbook materials and objects central to Christensen's case. While he is using them more as data on the past than as objects in still motion in the present, the drawings analyzed by Jordan relate to a breathtaking series of movements. As with Osage wedding clothing, the Kiowa drawings are predicated on materials obtained in the encounter with non-indigenous peoples. Similarly, they are not just carried from European factories to indigenous communities; they moved through complex social networks and were in turn made a part of a local material vernacular under conditions of cultural contact. In the Kiowa case, the Fort Marion drawings were made by prisoners forced half a continent away from their homes during a period of colonization, warfare, and forced assimilation. Today such drawings circulate in museum exhibitions and the private art market. In all the cases, people, ideas, raw materials, finished goods, and a great deal more are seen as in motion across social and geographic space even as the authors show the ways that these sometimes extralocal things are domesticated to local and personal interests, talents, and concerns.

Just as I do not wish to overly bound the series at its beginnings, I do not want to constrain your reading of the essays gathered here. Many more connections await to be drawn between them and between these contributions and other work being done across a range of fields. I know that the authors see themselves as contributing to conversations both already underway and heading in interesting new directions.

In conclusion, I would like to take this opportunity to address some specifics regarding the Material Vernaculars series as it has been framed at its start. As follows, the series mission statement opens by defining its scholarly scope and then characterizes the partners and means by which it will endeavor to pursue this work.

> The Material Vernaculars series presents ethnographic, historical, and comparative accounts of material and visual culture manifest in both the everyday and extraordinary lives of individuals and communities, nations, and networks. While advancing a venerable scholarly tradition

focused on the makers and users of handmade objects, the series also
addresses contemporary practices of mediation, refashioning, recy-
cling, assemblage, and collecting in global and local contexts. The
Indiana University Press (IUP) publishes the Material Vernaculars series
in partnership with the Mathers Museum of World Cultures at Indiana
University (IU). The series accommodates a diversity of types of work,
including catalogues and collections, studies, monographs, edited vol-
umes, and multimedia works. The series will pursue innovative publishing strategies intended to maximize access to published titles and will
advance works that take fullest advantage of the affordances provided
by digital technologies.

While evoked above, I wish to unpack these two parts of the series
mission briefly.

Reflecting the research focus and object collection of the Mathers
Museum of World Cultures, the series' concern with "ethnographic,
historical, and comparative accounts" sections off—as the museum
itself does—a portion of the wider field of material culture studies.
Archaeology, for instance, is a crucial contributor to this interdisci-
plinary field, but the series will not take on work in this part of the
field, although it is likely that scholarship published in the series will
be of value to archaeologists and series authors will, it is strongly
hoped, be conversant with the important work done in archaeology.
A wide range of disciplinary perspectives should find a comfortable
home within the series, but its disciplinary centers of gravity are, by
intention, the fields of folklore studies, social and cultural anthropology, ethnology, and cultural history. These are fields in which I work
and are those at the center of the museum's efforts. They are also
fields to which the IUP has long been especially devoted.

Within these fields, and as reflected in the series mission statement and my preceding remarks, Material Vernaculars will aim to
bridge the old and the new in the study of objects and images. The
old—but still vital—mode of material culture studies was closely associated with practical handmade objects of the sorts that preceded
lives saturated with commercially produced goods. In this frame,
material culture studies within folkloristic and anthropological ethnography contributed to the documentation of ancestral ways of life
that were disappearing in the face of colonization, modernization,
industrialization, and related processes of social, economic, and technological change. The renewed material culture studies of the 1990s

and 2000s instead took up new issues such as consumption and collecting that had been neglected by earlier scholarship. While enduring concerns such as with gift exchange and with craft economies transcended the old and the new eras in material culture studies, some "classic" concerns—such as the study of museum objects and collections or the close study of techniques of making—especially warrant renewed attention. It is my hope that the Material Vernaculars series will encourage and make available work that broadens and strengthens the field, including work that takes up classic concerns in innovative ways.

Similarly, the series provides an opportunity to encourage increased dialogue between scholars of material culture working within the neighboring—but not always mutually well-connected—fields of folklore studies, social/cultural anthropology, ethnology, and history. As a comparatively rare museum with programs bridging these fields, the Mathers Museum of World Cultures is an appropriate home for a book series aiming to connect them more meaningfully. In this regard, the series will hopefully parallel the accomplishments of the museum's open access, peer-reviewed journal *Museum Anthropology Review*. Since its founding in 2007, *Museum Anthropology Review* has published valuable scholarship across this same set of disciplines bringing them together to a degree that is otherwise unusual. What is hoped for by way of disciplinary dialogue can hopefully also become true for national and regional scholarly traditions. While centered in its own North American home context, the series—if successful—will promote scholarly engagement beyond the English-speaking North Atlantic.

One means of pursuing such goals is through a commitment to free and, where possible, open access. This brings us to the publishing approach that the series aims to explore and experiment with. Such experimentation is part of a larger agenda at the IUP and at IU as a whole. Now a unit of the IU Libraries and strongly supported by the university's provost, the IUP has a long and distinguished history as a scholarly publisher in the fields most relevant to this volume and to this series—cultural anthropology, folklore studies, ethnology, and history. Hosted and supported by a dynamic university with a strong commitment to innovation in scholarly communication, information technology, research librarianship, the social sciences, and the humanities, the IUP is undergoing a revitalization that situates it uniquely among North American university presses.

For the IUP, this series is intended to be part of a program of experimentation and innovation in scholarly publishing. Behind the scenes, this experimentation includes the development of new editorial workflows and production processes that aspire to accelerate the publication of significant scholarship, getting it into the hands of readers—both professional and lay audiences—more quickly and less expensively, without sacrifices in quality. In public view, the series is intended to pioneer innovations in access, making high-quality scholarship accessible for free via the internet while also making print and e-book editions, for those who prefer them, available in the established publishing marketplace.[1] For disciplines concerned with vernacular culture and with the lives of diverse global communities, improving access to scholarship is not just a desirable affordance of new digital technologies, it is an ethical obligation. Reflecting here on the chapters gathered in this volume, working-class Jews in Tel Aviv, Osage and Kiowa people in Oklahoma, retirees in Indiana, and scrapbook creators scattered around the United States all, for instance, have good reasons to expect access to this volume given that they or their communities figure in the chapters gathered here. Ethnographic books that draw on the goodwill and collaboration of individuals and groups scattered around the world simply cannot, as remains common today, be issued in print runs of one hundred copies and sold for one hundred dollars or more per copy. My hope is that Material Vernaculars will help develop sustainable and accessible publishing models that can become more and more widespread in the years ahead (Walters et al. 2015). In this aspiration, the book series, from a museum point of view, builds on the success of *Museum Anthropology Review*, which was similarly born out of a commitment to develop open access publishing strategies in partnership with the IU Libraries.

In order to suitably launch the new series, IUP Director Gary Dunham encouraged me to organize a Material Vernaculars event for the 2015 Annual Meeting of the American Folklore Society in Long Beach, California. A very successful conference panel was the result. In addition to providing a suitable opportunity to announce the series and to invite folklorists to contribute to it, the panel offered its audience four compelling investigations of material and visual culture. These significant contributions to the field are gathered in this collection. To them was added one additional study—Christensen's valuable chapter on contemporary scrapbooking as a material, visual, and discursive practice. While not in any fashion intending to delimit its boundaries, the studies presented here were chosen to suggest some of the diversity and scope that I hope will characterize the new Material Vernaculars book series that this collection inaugurates. I invite you to engage with these compelling studies and to consider extending the conversation further with your own work. *Museum Anthropology Review* always welcomes article-length works in the field of material culture studies and, with this launch, the Material Vernaculars series is available for those wishing to propose sole or joint-authored books as well as thematic edited collections, catalogues, and other longer-format works.

Acknowledgments

I express appreciation to all who have helped realize both this volume and the series of which it is a part. Thanks especially go to the volume's contributors, to the material culture scholars who generously strengthened our work through thoughtful service as peer reviewers, and to the staff at the IUP. It is a rare and wonderful thing to work on this volume and series with outstanding publishing professionals who themselves are also talented students of material culture, but such is our good fortune in working with folklorist and Acquisition Editor Janice Frisch and archaeologist and Press Director Gary Dunham. Their commitment to publishing new scholarship exploring the richness and complexity of the human condition and to doing so in innovative ways is inspirational.

Note

1. Readers of this introduction who know something of my work over the past decade will, I hope, see this aspect of the series as an extension of my activist and scholarly work on open access publishing, both in journals and, more recently, working toward frameworks for open-access monograph publishing (Jackson 2012; Jackson and Anderson 2014; Kelty et al. 2008).

References

Bell, Joshua A. 2010. "Out of the Mouths of Crocodiles: Eliciting Histories in Photographs and String-Figures." *History and Anthropology* 21: 351–73.

Berger, Arthur Asa. 2009. *What Objects Mean: An Introduction to Material Culture.* Walnut Creek, CA: Left Coast Press.

Geismar, Haidy. 2011. "'Material Culture Studies' and other Ways to Theorize Objects: A Primer to a Regional Debate." *Comparative Studies in Society and History* 53: 210–18.

Glassie, Henry. 1999. *Material Culture.* Bloomington: Indiana University Press.

Jackson, Jason Baird. 2012. "We Are the One Percent: Open Access in the Era of Occupy Wall Street." *Anthropologies.* March 1. http://www.anthropologiespro ject.org/2012/03/we-are-one-percent-open-access-in-era.html.

Jackson, Jason Baird, and Ryan Anderson. 2014. "Anthropology and Open Access." *Cultural Anthropology* 29: 236–63. doi:10.14506/ca29.2.04.

Kelty, Christopher M., Michael M. J. Fischer, Alex "Rex" Golub, Jason Baird Jackson, Kimberly Christen, Michael F. Brown, and Tom Boellstorff. 2008. "Anthropology of/in Circulation: The Future of Open Access and Scholarly Societies." *Cultural Anthropology* 23: 559–88. doi: 10.1111/j.1548-1360.2008.00018.x.

Kemble, Amanda, Briel Kobak, Joshua A. Bell, and Joel Kuipers. 2015. "A Day in the Life of a Cell Phone Repair Technician in the Digital Age." In *A World of Work: Imagined Manuals for Real Jobs,* edited by Ilana Gershorn, 179–93. Ithaca, NY: Cornell University Press.

Löfgren, Orvar. 2012. "Material Culture." In *A Companion to Folklore,* edited by Regina F. Bendix and Galit Hasan-Rokem, 169–83. Malden, MA: Wiley-Blackwell.

Miller, Daniel. 2010. *Stuff.* Malden, MA: Polity Press.

Shukla, Pravina. 2008. *Grace of Four Moons: Dress, Adornment, and the Art of the Body in Modern India.* Bloomington: Indiana University Press.

Tilley, Chris, Webb Keane, Susanne Küchler, Mike Rowlands, and Patricia Spyer, eds. 2006. *Handbook of Material Culture.* Thousand Oaks, CA: Sage.

Walters, Carolyn, James Hilton, Jason Baird Jackson, Scott Smart, Nick Fitzgerald, Gary Dunham, Shayna Pekala, Paul Courant, Sidonie Smith, Meredith Kahn, Charles Watkinson, Jim Ottaviani, and Aaron McCollough 2015. *A Study of Direct Author Subvention for Publishing Humanities Books at Two Universities: A Report to the Andrew W. Mellon Foundation by Indiana University and University of Michigan.* Bloomington: Indiana University and University of Michigan. https://scholarworks.iu.edu/dspace/handle/2022/20408.

JASON BAIRD JACKSON is Director of the Mathers Museum of World Cultures and Professor of Folklore Studies in the Department of Folklore and Ethnomusicology at Indiana University.

Gabrielle A. Berlinger

1   Searching for Home in the Ephemeral Architecture of the *Sukkah*

WHAT IS THE relationship between materiality and spirituality in Jewish life? When and how do material things and Jewish religious experience meet and affect one another? And to what end? These questions crystallized for me during a sixteen-month research project that combined my two fields of interest: ritual practice and vernacular architecture—the study of common structures and everyday landscapes. For a period of eight years that culminated in this project, I conducted ethnographic fieldwork around the ancient Jewish festival of *Sukkot*, an annual commemoration of the Israelites' forty-year journey through the Sinai Desert after the exodus from Egypt. I chose to study this holiday because of the temporary constructions that characterize its religious observance—ritual shelters called *sukkot* (Hebrew: tabernacle; booth) (Fig. 1.1). For seven festival days each fall, observant Jews around the world design and build these outdoor ritual architectures that evoke a multiplicity of memories and meanings, primary among them the impermanent shelters that the ancient Israelites used during their search for the Promised Land.[1] My study explored the role of material culture in the contexts of migration and religious expression.

What follows are four scenarios in which contemporary *sukkah* (singular of *sukkot)* builders and users express distinct and intentional notions of "home" through the material architecture of their sukkot. Because the sukkah is a structure symbolic of the domestic space, its material manifestation often communicates the builder's notion of "home." My documentation of current-day Sukkot observance in Bloomington, Indiana; Jerusalem, Israel; Brooklyn, New York; and

FIGURE 1.1.
The *sukkah* (Hebrew: tabernacle; booth) is the ritual structure that observers of
the annual Jewish festival of *Sukkot* build and use for the weeklong holiday. This
typical sukkah was constructed using green plastic tarps as walls and palm tree
branches as roof covering (South Tel Aviv, Israel, 2010).

Tel Aviv, Israel illustrates how Jews construct ritual structures that
convey personal and collective histories, conditions, values, and beliefs. Significantly, however, each individual locates meaning in a
different material element of the sukkah's construction: the frame
of the sukkah, the interior decoration, the roof covering, or the total-
ity of the architecture. This diversity of interpretation of the ritual
structure demonstrates the range of meaning attached to the materials of Jewish ritual practice. Such variation drives the theoretical and
methodological approach to what Leonard Primiano calls "vernacular religion," or "religion as it is lived: as human beings encounter,
understand, interpret, and practice it" (Primiano 1995, 51). Rather
than rely on the often hierarchical and reductive categories of "folk,"

"popular," "unofficial," and "official" to describe religious practice, Primiano identifies "vernacular" as a term that effectively draws on the perspectives of religious and folklore studies to emphasize the personal and private, and the artistry and agency that characterize individual belief systems. Primiano's theory and approach to "vernacular religion" inform my study of Sukkot observance as I examine the diverse contexts and expressions of Jewish belief.

As such, the following four interpretations of the sukkah's material form communicate four notions of "home": home as nation, home as ethnicity, home as the Divine, and home as a just society. As French philosopher Gaston Bachelard has observed, "the house shelters daydreaming, the house protects the dreamer, the house allows one to dream in peace" (Bachelard 1994 [1958], 6). In the context of Sukkot, the dream to which Bachelard refers is the notion of "home" that each sukkah builder nurtures through her or his practice—an ideal state of consciousness in which thoughts, memories, and hopes come together.

The written origins of Sukkot observance appear in Leviticus, the third book of the Hebrew Bible, where in verse 23:42–43, God commands Moses, "You shall dwell in booths for seven days. All native Israelites shall dwell in booths, that your generations may know that I made the people of Israel dwell in booths when I brought them out of the land of Egypt" (ESV). Throughout history, religious, academic, and legal authorities have produced lengthy interpretations of this brief commandment that prescribes the observance of Sukkot. The essential requirements of sukkah construction, however, are few. According to *halakha* (Hebrew: Jewish religious law), first, the sukkah must have at least two full walls that are connected to each other, and a third wall at least one *tefach* (Hebrew: handbreadth) wide; and, second, the *schach* (Hebrew: roof covering) must be made of organic matter gathered from the earth (Fig. 1.2). No prescriptions about decoration of the sukkah exist in the original commandment (Fig. 1.3). Observant Jews pray, eat, socialize, and even sleep inside these structures for the week of the holiday, recalling the Israelites' historic search for home and homeland, and raising consciousness about such concepts as human survival, connection to nature, homelessness and home, exclusion and belonging, and materiality and spirituality.

In this study of sukkah construction as an expression of "home," the notion of belonging resonates with particular significance. First, one must know that the practice of hospitality lies at the heart of Sukkot observance, performed in the custom of welcoming outsiders into one's sukkah for food, drink, and rest for the duration of the festival. Observers of this custom frequently recall the biblical story of Abraham welcoming three strangers into his tent (Genesis 18) as the model for their generosity. In return for his kind gesture, Abraham's unexpected guests reveal to him that he and his wife will soon expect a child. Hospitality to strangers, it is believed, opens a door to unknown fortune. As one sukkah builder explained to me, "God loves hospitality. Abraham our Father searched for guests to host, for God said first you will respect hospitality, then you will respect me. God loves a person who welcomes guests into his home." Hospitality grounds social behavior for the week of Sukkot, both affirming and challenging presumed positions of "insider" and "outsider." As a temporary ritual structure defined by its "moment in and out of time," the sukkah

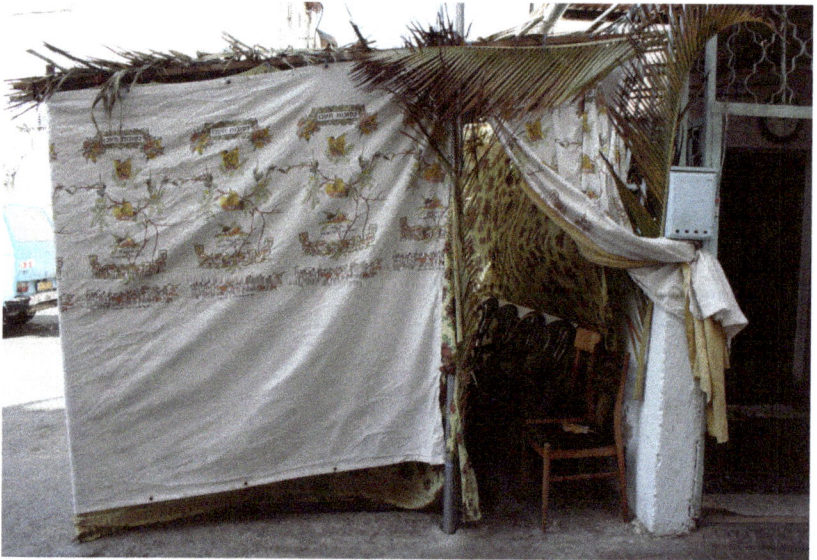

FIGURE 1.2.
Two biblical prescriptions determine the construction of a sukkah: it must have at least two full walls that are connected to each other and a third wall at least one handbreadth wide; the roof covering must be made of organic matter gathered from the earth (South Tel Aviv, Israel, 2010).

FIGURE 1.3.
Although no prescriptions about decoration of the sukkah exist in the original biblical commandment, many individuals aesthetically customize the sukkah's interior to reflect personal values and cultural conditions. This sukkah, decorated with tapestries, photographs, and hanging tinsel chains, was built by a Bukharan Jewish family from Uzbekistan (South Tel Aviv, Israel, 2010).

creates a space of inclusivity, a liminality that suspends ordinary boundaries and beliefs about belonging until the holiday ends (Turner 1969, 96). In this way, the commandment to welcome the stranger erases the division between outsider and insider and affirms the fact of belonging.

Given this custom of hospitality in the context of Sukkot, the notion of belonging is particularly heightened in the Jewish Diaspora. Historically, the Jewish Diaspora refers to the "global 'scattering' of the Jewish people that took place in the years following the Babylonian captivity (sixth century BCE) and especially after the destruction of the temple by the Romans in 70 CE"—a fact that has engendered a "Jewish self-understanding of collective peoplehood in exile" (Jackson 2006, 18–19). The consciousness of this condition was perceptible in my focus on Sukkot ritual performance, whether in a sparse Jewish population of a small, midwestern American city, a dense, multicultural population of Jewish immigrants in an Israeli city, or an insular Orthodox Jewish population in a large East Coast American city. In these contrasting social environments, discussions about build-ers' choices regarding sukkah construction and decoration revealed the interlocking local, national, and international networks to which they felt they belonged. American Jews in the Midwest might decorate their sukkah with an Israeli flag to express a relationship with Israel, and beside it, a treasured Wedgwood plate passed down in their family. Bukharan Jews who emigrated from Uzbekistan to Israel might hang a picture of an Uzbeki politician whom they still support alongside fresh pomegranates and dates, two of the "Seven Species" connected to the Land of Israel in the Hebrew Bible. And American Jews in the Northeast might attach laminated drawings of the Western Wall in Jerusalem to the sukkah's interior while dangling above them are tinsel decorations and strings of light purchased in the "Christmas decora-tion" aisle at the local craft supply store. For Jews who perceive them-selves as part of the Diaspora, such material interventions in the space of the sukkah express distinct senses of belonging—in this case, to American, Bukharan, and Israeli cultures. The diverse materiality in the sukkah tradition thus offers salient examples of "insider"-"outsider" identification, weaving together near and distant societies and land-scapes. As Jason Baird Jackson has observed, "the artistic, expressive, and customary practices of globally dispersed populations—which often take on the privileged and self-conscious status of 'heritage'—are central to the establishment and maintenance of a diasporic identity" (Jackson 2006, 19). The particular material traditions of sukkah con-struction and decoration illustrate how this expression of Jewish "her-itage" helps to define a diasporic identity in which a sense of belonging is achieved through a nuanced, individual orientation of the "Self" to the "Other."[2]

This wish for belonging that resonates with Sukkot observance is heightened in a second way in the context of the Jewish Diaspora—this time, through the observers' reflection on attaining a just society. During the week of Sukkot, observers move out of their permanent homes to dwell outdoors, in impermanent shelters. One sukkah builder interpreted for me the significance of placing the sukkah outside: "The sukkah is a space of meeting. It's supposed to be a way to be together, in solidarity and partnership, before the arrival," he said, referring to the journey to the Promised Land in the biblical narrative. For him, the neutral space of the outdoor sukkah was a place to pause and be together, a moment to cross social and physical boundaries. The ritual of hosting and being hosted in the sukkah prompted this builder each year to reconsider his own boundaries, and to open his home and heart to unfamiliarity and difference. "It's not the idea of *v'ahavta l'reaacha camocha* ('Love your neighbor as you love yourself'), which is very important in and of itself," he continued, "but how do people on a long journey in the desert arrive in a new land and all live together? How do people of different cultures come together to create one society?" The temporary sukkah has the power to bridge differences by allowing for reflection on how to live with each other, in fairness and in peace. Interpreting the space of the sukkah as a biblical moment of potential that preceded the Israelites' arrival in the Promised Land, this builder reflected deeply on how to create a society of equals today.

Sukkot, a holiday that recalls the shelter provided in the desert during the Israelites' search for "home," holds at its core the yearning to belong, and the acceptance of others as equals. The following four case studies illustrate how individuals both in the "Diaspora" and in the "Promised Land" manifest their particular ways of belonging through the construction of "home."

Contemporary Practice

My fieldwork journey begins in Bloomington, Indiana, a midwestern university town, in 2007. There, we meet Bakol Geller, an actress and teacher who grew up in Canada and lived in Israel before moving to Indiana. Bakol experienced little formal Jewish practice growing up and attended both the Jewish renewal and conservative services at Bloomington's synagogue, explaining that she moved between categories of denominational affiliation as they fit her practice, which she described as "eclectic Judaism."

Using the back of her house to support one of the walls of her sukkah, Bakol built her structure, paradoxically, by her clear blue swimming pool (Figs. 1.4, 1.5). The frame of her sukkah is built out of purchased lumber, metal brackets, green tarps, and gathered brush reused each year (except for the brush). In her construction, Bakol adheres to *halakhic* prescriptions and explains her strict observance as a way of connecting with "something bigger" than her own life experience: "It's very real to me that even when I'm gone, something that

FIGURE 1.4.
Bakol Geller in front of her sukkah (Bloomington, Indiana, 2007).

FIGURE 1.5.
Bakol Geller builds her sukkah by the pool behind her house (Bloomington, Indiana, 2007).

I belong to will go on that was always here, and that will always be here."
Bakol follows the religious rules of construction as a way in which to
"step into a stream" of Jewish practice.

Bakol's Jewish identity is defined in part by adhering to the religious prescriptions written in Jewish texts, but it is also defined by her
connection to Israel, actualized in her ritual process. "I know [this
observance] is part of the Jewish religious practice but so much of
that for me is tied into being in Israel," she said. "In Israel we lived
more simply, in smaller quarters. . . . I like the idea of [the sukkah] as
a reminder that our refuge isn't in the material. . . . That what's going
to happen doesn't depend on how thick the walls are. It's about being
outside and about being in nature," she said. Bakol's sukkah construction and use reawaken her experience in Israel, which is central
to her current sense of self.

Beyond the frame of her structure, Bakol creates meaning through
her personalized array of ornaments inside—pictures cut from old
Jewish calendars, Israeli flags, and family heirlooms that evoke memories of her family and of Jewish culture: "It always seemed like a desert
thing, a Jewish thing, to have a tambourine. . . . And I always hang
Israeli flags here because that's part of what is being Jewish in America
for me. . . . And when my mother died, I didn't take much but *Shabbos*
candles and this little Wedgwood plate. I bring objects that are symbolic of my identity," Bakol concluded (Fig. 1.6). Bakol finds as much
meaning in the objects with which she adorns her sukkah as in its
construction and tells me that an undecorated sukkah would be "a
sukkah that hasn't been finished." Although not prescribed in the
Torah, many who decorate their sukkah cite a verse in Exodus
(15:2) as reason to beautify their ritual structures. This verse, which
reads "This is my God and I will adorn Him," has inspired Talmudic
interpretations that nurtured the Jewish principle of *hiddur mitzvah*—
the aesthetic enhancement of a *mitzvah* (Hebrew: commandment).
This principle holds that the aesthetic embellishment of any expression of devotion enhances the act by appealing strongly to more of
the five physical senses.

In her study of Mexican American women's home altars, folklorist
Kay Turner discovers meaning in object assembly. She characterizes
the altars that she studies as having an "aesthetic of relationship" that
is based on "images and objects that have no immediate affinity [but]
are nonetheless yoked together to forge new, interrelated meanings"

FIGURE 1.6.
Bakol Geller decorates the interior of her sukkah with laminated pages from Jewish calendars, Israeli flags, and family heirlooms that evoke memories of her family and historic Jewish experience (Bloomington, Indiana, 2007).

(Turner 1999, 95, 98). The meaning that Turner finds in structural relation rather than in solitary form emerges in Bakol's ritual decoration. Her assembly of objects binds together her individual Jewish identity with a collective Jewish history and nation.

From the Midwest, we now move to the Middle East, where for sixteen months between 2010 and 2011, I based myself in Southeast Tel Aviv, Israel, to document Sukkot observance in a dense, diverse community of Jews from Middle Eastern, North African, and Central Asian countries. Although I concentrated my time in this urban quarter, I traveled to neighboring cities and towns during the holiday for comparative context. Each year, I made one such trip to Jerusalem to visit Drori Yehoshua, a man of Kurdish descent. Drori was born in Jerusalem and worked as Rosh Beit Midrash, or Head of House of Study, at Memizrach Shemesh, "a Beit Midrash (House of Study) and Center for Jewish Social Activism and Leadership in Israel."

When you enter Drori's home and walk through the kitchen and bedroom to the backdoor, you come upon a staircase that leads you down to his backyard. There, he builds his sukkah each year with a metal frame six meters wide, three meters long, and two meters tall (Fig. 1.7). The frame is a standard size according to the manufactured dimensions, but he has pieced this frame together from different poles that previously belonged to his brother and to his grandfather (Fig. 1.8). I ask Drori how he chose these walls for his structure, and he

FIGURE 1.7.
Bird's eye view of Drori Yehoshua's sukkah that he builds each year in his backyard (Jerusalem, 2011).

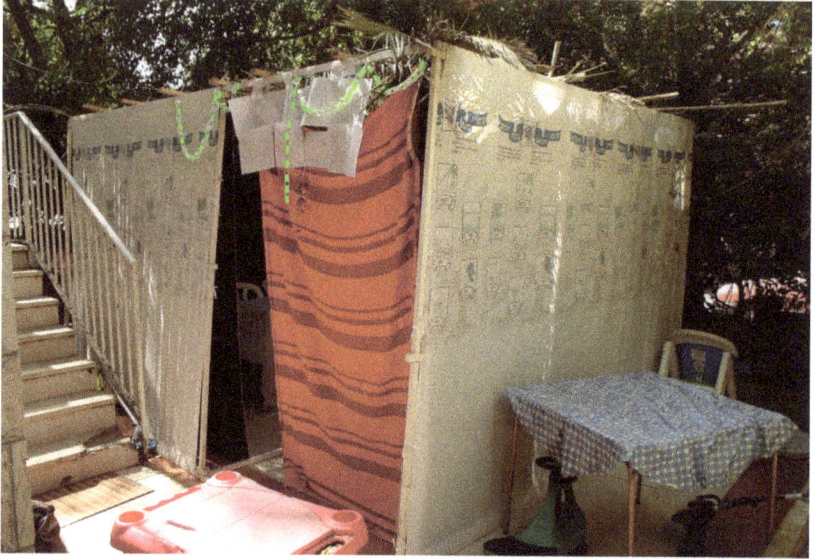

FIGURE 1.8.
Front view of Drori Yehoshua's sukkah, built with a metal pole frame, fabric walls, and gathered brush for the roof (Jerusalem, 2011).

dwells on the description, weaving together his own life story with that of his family's history to explain the sukkah's design. "These are the *parochot* [Hebrew: ornamental curtains that hang in front of the holy ark in a synagogue] from the *aron hakodesh* [Hebrew: holy ark that contains the Torah scrolls] of two Kurdish synagogues here in Jerusalem— Barashi and Bamedi—which represent two villages in Kurdistan. . . . My father would pray in Barashi, where I pray today. My mother would pray in Bamedi. These synagogues are opposite each other" (Fig. 1.9). The parochot that Drori describes and that hang before us are beautifully woven tapestries made of maroon velvet and golden thread; green, red, and yellow bouquets of flowers embroidered with golden Hebrew words; filigree patterns; and Jewish stars, crowns, and Decalogues. They are old and visibly used (Fig. 1.10).

As we examined them, Drori's younger son entered the sukkah, pointed to the curtain that faced us, and announced, "This one was hanging behind Grandma and Grandpa when they got married." Drori then tells me that his father Moshe immigrated to Israel from Kurdistan with his family in 1952. His mother Margolit was born in Israel, her family having emigrated from Kurdistan in 1928.

FIGURE 1.9.
Interior view of Drori Yehoshua's sukkah. Each Sukkot, Drori borrows eight
parochot (Hebrew: the curtains that cover the holy ark in a synagogue) from local
Kurdish Jewish synagogues for the walls of his sukkah (Jerusalem, 2011).

FIGURE 1.10.
Far interior view of Drori Yehoshua's sukkah. A Kurdish coffeepot, the Hebrew
Bible, and a framed photograph of Drori's parents stand on the table (Jerusalem,
Israel, 2011).

He smiled and said, "They married in Israel, a Kurdish wedding," and then reached for a framed black-and-white photograph that had been standing upright on the end of the table (Fig. 1.11). Pointing to the blurry backdrop behind the young couple in the photograph with one hand and pointing to the curtain that his son had just identified with the other, Drori said, "That's the same curtain! That's their wedding, and there they are standing in front of it."

"So how do you have these parochot?" I ask him. Drori replies that he borrowed them from his local synagogues and explains:

> These parochot are interesting because the older they are, the less frequently they are taken out of storage, and this one [he points to one hanging behind us] hasn't been out for 15 years. In my opinion, it is one of the most beautiful. This one also doesn't get taken out. This is a problem. So I take these out for Sukkot, and hang them, use them, and sometimes if I see a tear, I sew it, or clean it. These are very, very important. Most times, parochot are made in memory of someone.

Drori's relationship with these Kurdish synagogues and their social communities is mutually nourishing. He serves as the community's

FIGURE 1.11.
A close-up of the photograph of Drori Yehoshua's parents, Margolit and Moshe, on their wedding day. In the photograph, they stand in front of the same *parochet* that now hangs in Drori's sukkah.

cantor, and the community serves him by lending him these parochot during Sukkot. His material construction is a social reconstruction of his heritage and an assertion of his Kurdish ethnicity.

All of the walls of Drori's sukkah are made from parochot: four that he borrowed from his father's synagogue and four that he borrowed from his mother's synagogue. Equally representing the two sides of his family, they not only track his parents' roots through these materials of spiritual and cultural devotion, but they ground Drori in his past and present life as well. "In my house, I have things I bought from IKEA, from here and there, and maybe I have a coffeepot from Kurdistan, but it's a little hard to tell who I am from just that. When I go outside from my house, there are possibilities in the sukkah to create a home design that says who I am. So I hang the parochot. This is my father, and this is my mother," said Drori, looking between the two sides of his sukkah. "What more does a man need to say than where his father and his mother are from? It's a kind of identity."

These tapestries represent Drori's Kurdish ancestral roots, although they were made and used in Israel. He thus connects various "homes" through their use during Sukkot: the Israelites' temporary shelter, the sense of home that is embodied in the memory of his parents, the house in Jerusalem behind which he builds this sukkah, and the two Kurdish synagogues, or literally "houses of prayer" that bridge the communities from where his family emigrated, with the Kurdish communities in Jerusalem where he was born and continues to live. By building walls of meaning around his family during Sukkot, Drori consolidates all of these notions of home for an annual, ephemeral experience of personal placement.

Our third destination is Brooklyn, New York, where during Sukkot of 2014 and 2015, I documented holiday observance within the Lubavitcher Jewish community (Chabad), a Hasidic movement of Orthodox Judaism established in Russia in the mid-1700s. When you emerge from the number 3 subway station at Kingston Avenue in Crown Heights, you face 770 Eastern Parkway, the Chabad headquarters and home of the movement's former Rebbe—or spiritual leader—and thus, also, the social and spiritual center of the sect (Fig. 1.12). Inside this building, now a pilgrimage site for Chabad Jews from around the world, Rabbi Chaim Halberstam established and runs the Lubavitcher Communication Center from a small room

FIGURE 1.12.
Front view of 770 Eastern Parkway, the Lubavitch world headquarters (Crown Heights, Brooklyn, New York, 2006).

in the back—a control room out of which all of the Rebbe's sermons and teachings have been recorded and are disseminated.

Halberstam was born in Israel and immigrated to New York as a young man. He lives several blocks from 770 Eastern Parkway on a quiet street lined with old trees and brick houses fronted by terraces and stoops. When I visited him the day before Sukkot, he walked me down the side alley to the back of his house where he builds his sukkah out of plywood and repurposed metal poles (Fig. 1.13). "If you look at my makeshift sukkah," he says, "people wonder how it's held up. The whole sukkah is held with eight bolts." Halberstam painted large numbers on the interior wooden panel walls to remind himself how to assemble the structure each year, and the roof covering is an assortment of bamboo mats that he annually reuses (Fig. 1.14). The structure is bare but for a fluorescent light hanging from the roof inside. When I ask him what changes are still to be made before the holiday begins, he says it's done—nothing is to be added but a table and chairs. "You know Chabad doesn't have any

FIGURE 1.13.
Rabbi Chaim Halberstam stands with his wife Mindy in front of the sukkah that they build behind their house. He constructs the sukkah out of metal poles and plywood panels, all held together with eight bolts (Crown Heights, Brooklyn, New York, 2015).

decorations," he says; "Chabad is different than the whole world in two ways. First, no decorations. And second, when it rains, the whole world goes into the house to eat. Chabad will stay in the sukkah and eat in the rain. I once started eating a soup, and I kept eating and eating and never finished because it kept raining," he says with a straight face and then a smile. Rather than decorations, the material that deserves attention, says Halberstam, is the schach (Fig. 1.15). The Talmudic interpretation that the roof covering should not exceed a height of approximately thirty feet, he says, is derived from the need to see the schach when you enter the sukkah. "The schach has to be noticed," he says; "Your attention should be on the schach, not on all the beautiful drawings and decorations. Our custom is even that when you make the special *brachah* (Hebrew: prayer) for the sukkah—*L'shev b'sukkah*—we look up at the schach. We should feel the sukkah." The schach, revealing the stars of the night sky through its weave and shading the hot day's sun with the weave, is religiously interpreted as the "Clouds of Glory" with which God surrounded the Israelites to protect and comfort them in the wilderness (Rubenstein 1994).

FIGURE 1.14.
Rabbi Chaim Halberstam paints large white numbers on each plywood panel, so that he remembers the order in which to construct the frame. No decorations adorn the interior, as is custom in Lubavitch Sukkot observance, but a table, chairs, and lamp will furnish the inside (Crown Heights, Brooklyn, New York, 2015).

Two days later at lunch in the sukkah of another Chabad family a few blocks from Halberstam, I observed the men present raise their eyes to the woven brush as they recited the prayer before we began our meal. This sukkah belonged to the family of Mayer Preger, the relatives of whom live throughout Crown Heights. Explaining the Chabad custom of not decorating the sukkah to me, Mayer said, "No one took out the decorations or said don't focus on it because they didn't appreciate it. It's really nice and that's the problem. You're making it very beautiful and then appreciating just the beauty of it, getting distracted by the superficiality of it. Chabad philosophy as a whole is that we try not to be distracted by the mundane physicality of it. So do I appreciate decorations? Yes. When I go into someone's sukkah that's decorated, do I like it? Of course! But I also appreciate what our custom is trying to do."

FIGURE 1.15.
Rabbi Chaim Halberstam uses a *schach*, or roof covering, made out of bamboo mats (Crown Heights, Brooklyn, New York, 2015).

I gained a third perspective on Chabad's interpretation of the sukkah's structure from Rabbi Ephraim Piekarski, a former headmaster at Educational Institute Oholei Torah, the main *yeshiva* (Hebrew: Jewish school for religious instruction) in Crown Heights. When I asked Rabbi Ephraim Piekarski about Chabad's restraint in decorating the sukkah, he replied: "We strike out for the decorations, but it has to be put in context." Chabad philosophy distinguishes its worldview from that of classical Judaism, he said. "Doing a mitzvah is a good thing. I'm doing what God wants, it's a good feeling, God gave me a beautiful holiday, He cares about us, and if we observe it properly, He'll take care of us—it's a warm and fuzzy feeling. This is classical Judaism," he said; "But for us, a mitzvah is more than just a beautiful thing which you can explain. The beauty of all sitting together and the unity and all that stuff that's beautiful about Sukkos, that's true," he continued, "but there's something much deeper. . . . We have this intrinsic feeling that a mitzvah is a strong bond with *HaShem* (Hebrew: God)." In the Chabad worldview, the roof covering is the material of meaning because God commanded that material, and only that material. By focusing only on this single commandment, Lubavitcher Jews transcend their material world through the spirit of the schach. As Mayer Preger had concluded, "The whole point is that our physical life and our spiritual life should not be two separate lives, they should be one."

The three cases above illustrate individual interpretations of the sukkah. Personal history and current circumstance motivate these three individuals to dream of connections that make them feel "at home" in their temporary structures, and through their specific material expression, they bond with a homeland, an ethnicity, and a God. My fourth and final case takes us back to Israel, to the main site of my field research in South Tel Aviv where an entire community, as opposed to an individual, seeks "home" in the sukkah, and where the entire structure, rather than a single material element of its construction, enables these individuals to dream.

For sixteen months between 2010 and 2011, I conducted fieldwork in Shchunat Hatikva, a low-income neighborhood of South Tel Aviv, Israel. Shchunat Hatikva, translated as "Neighborhood of Hope," is a multiethnic quarter of the city that received waves of Jews from Yemen as early as the 1930s and waves of Iraqi and Iranian Jews by the 1950s. Today, it is also home to Jews from a variety of Middle Eastern, North African, and Central Asian countries, including Syria, Egypt, Libya, Morocco, and Uzbekistan. Although I chose to study Sukkot observance in this community to examine the role of this "home-making" ritual in the formation or fragmentation of a multiethnic society, I did not anticipate that two sociopolitical events would occur during the period of my research that would add new layers of meaning to the contemporary observance of Sukkot. As noted at the start of this essay, the construction of a temporary home, symbolic and literal, creates a physical, intellectual, social, and ethical space in which to explore the notion of belonging—in particular because the central custom of Sukkot is the welcoming of strangers into one's sukkah. In this fourth case study, this custom of hospitality was challenged by the two unexpected developments that occurred in 2010 and 2011. In the context of worldwide migrations that marked those years and are still underway, the ritual construction of the sukkah was put into stark relief as a relevant meditation on the meaning of "home" for the millions who seek it.

In 2005, a new wave of non-Jewish foreigners moved into South Tel Aviv. Sudanese and Eritrean asylum seekers, fleeing violence and hardship in their home countries, sought refuge in Israel and were placed in low-income neighborhoods such as Shchunat Hatikva. The population crossing into Israel reached its peak during the period of my research, and by 2013, the number of Eritrean and Sudanese asylum seekers living within Israel's borders had exceeded sixty thousand.[3] These new populations occupy a liminal place in Israeli society.

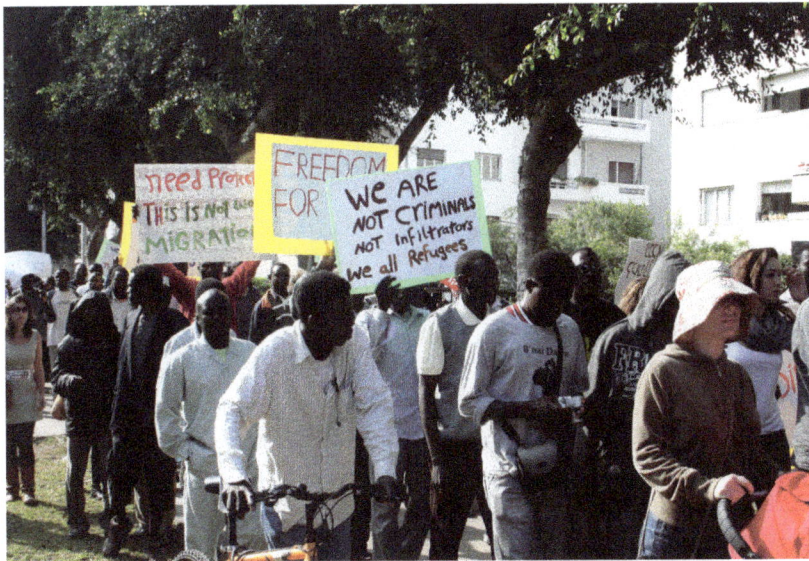

FIGURE 1.16.
Eritrean and Sudanese asylum seekers, tens of thousands of whom were placed in low-income areas of Israel such as South Tel Aviv, protest in central Tel Aviv for refugee rights and protection (Tel Aviv, Israel, 2011).

Allowed to enter the country in search of sanctuary and a better life, they nonetheless have unrecognized status, are denied work permits and subsist on insufficient government aid (Fig. 1.16). The public outcry by the Hatikva community regarding the asylum seekers provided me with insights into not only how veteran Jewish residents viewed newcomers to their neighborhood, but how they viewed themselves, more privileged Israelis, and the government—all critical perspectives on the notions of Self and Other that informed my understanding of their religious, social, and cultural practices. They blamed the government for placing thousands of desperate individuals in the nation's poorest areas, as reports of increased crime in the neighborhood circulated and concerns for its Jewish identity grew— the absorption of large numbers of non-Jews with different social, cultural, religious, and linguistic traditions seeming to them to threaten it.

Two contrasting yet parallel searches for home now coexist in the same place, provoking intense questions of social belonging—the first by a Jewish population historically treated as Other in Israel, ritually reliving the dislocation of Sukkot, and the second, by a non-Jewish population, currently treated as Other in Israel, seeking refuge. I was not alone in recognizing this parallel in the situations of the two populations. As I conducted research interviews about Sukkot observance, various individuals noted the similarities between the ancient narrative of Sukkot and the contemporary plight of the asylum seekers. They perceived the tension between the ritual obligation to welcome the stranger into one's home and the fear of literally doing so. Dror Kahalani, a resident of Hatikva, articulated the contradiction between empathy and resistance: "God told us to remember that you too were in Egypt. Don't forget that you were foreigners. . . . We need to remember that our fathers were also [outsiders] in a foreign country."[4] But, like several others who spoke about the tensions in the neighborhood, he added: "It's written in the Talmud: 'Your own city's poor before the poor of another city.'"[5] Because of the government's historic and current neglect of its own long-standing poor communities, veteran residents of Hatikva felt the need to place their own needs over the needs of others.

The first half of the time that I spent in Hatikva was framed by this backlash against the thousands of asylum seekers settling in South Tel Aviv. How does a community reconcile the contradiction between an ethical and spiritual commitment to inclusion, reinforced by a historic experience of exclusion, and the need to compete for diminishing resources heightened by the fear of losing one's cultural and religious identity? This conflict between Self and Other, at the heart of the Sukkot observance that I documented in Hatikva, resonates with migrants and immigrants all over the world who have formed, and who continue to form, new constituencies in new places, and with the established communities who receive them. Under the pressure of demographic changes, notions of Self and Other are intensified and values and behaviors begin to change. For whom do we open the doors of our homes and our nation? When, and why? And how do we include them in our economic, political, cultural, and social societies?

The second unanticipated event that occurred during my fieldwork period was the Israeli social protests that erupted in the summer of 2011. These protests were unprecedented in the nation's memory because of the shift in focus from issues of national security to issues within civil society. The protests also played a meaningful role internationally, erupting after the waves of social demonstrations set off in 2010 by the Arab Spring and before the Occupy Wall Street movement of 2011 that fomented demonstrations in more than one hundred American cities. In Israel, beginning in July 2011, social protests centered on the lack of adequate, affordable housing grew, and when the week of Sukkot arrived in October during the demonstrations, long-suffering low-income communities erected sukkot in their protest encampments in Hatikva's public park (Fig. 1.17). "Occupy Judaism" and "Occupy Sukkot" became widespread slogans during the holiday throughout the world, and the resonance between the ancient holiday and the current situation in the country was unmistakable. The sukkot that were built in the protest camps both evoked and contested the holiday's historic and religious meanings as the temporary lack of shelter at the core of the Sukkot narrative was confronted by the permanent lack of shelter in the protesters' actual lives. In New York, Dan Sieradski, organizer of Occupy Judaism NYC, declared, "There is no better place to celebrate the festival of Sukkot this year than right here at Occupy Wall Street. We stand in solidarity with all those who are challenging the inequitable distribution of resources in our country, who dare to dream of a more just and compassionate society" (Fleischman 2011).

FIGURE 1.17.
Demonstrators in South Tel Aviv erected sukkot in Gan HaTikva, the public park where they pitched and inhabited tents for over three months in protest of poor public housing conditions in low-income areas of Tel Aviv. Here, a small, prefabricated sukkah stands before a large protest tent in which demonstrators slept (South Tel Aviv, Israel, 2011).

As a dynamic tradition that generates reinterpretation through annual performance, the construction of the sukkah accrues new value in each context of use. The crisis of the asylum seekers and the protest movement that framed my research in Israel vividly demonstrated the experiences of dislocation and economic privation among various populations, raising questions about the relationship between ideal notions of "home" and the realities of inadequate housing in the construction of a stable society. Both of these circumstances also provoked a reconsideration of the notion of social belonging through their search for shelter, by foreigners in Israel and by Israeli citizens themselves. The ancient narrative of the holiday, its contemporary ritual commemoration, and current-day conditions of living continue to resonate strongly.

In this last case study, the full form of the sukkah is the material of transformation. An architecture of equality, the sukkah is a symbol of permanent shelter, an evocation of settlement, and a hope for security. "The sukkah is similar to a house, but the house is temporary," one Iranian Jewish resident of Hatikva told me; "[the house] comes and goes, while the sukkah is forever." In 2010, the sukkah's symbolism challenged South Tel Aviv's veteran residents to question whether the asylum seekers belonged in their homes or not, and in 2011, it illuminated the lack of permanent housing in their own society. As an instrument through which to dream, the sukkah enabled deprived peoples and struggling communities in South Tel Aviv to imagine an ideal society marked by justice and equality.

Conclusion

Philosopher Aviezer Tucker theorizes the concept of home as a condition that allows for personal self-fulfillment. He writes that "home is the reflection of our subjectivity in the world" (Tucker 1994, 184). Drawing on the ideas of Czech philosopher Vaclav Havel who regards the "home" to be "an existential experience that can be compared to a set of concentric circles on various levels, from the house, the village or town, the family, the social environment, the professional environment, to the nation including culture and language (Czech or Slovak), the civic society (Czechoslovak), the civilisation (European), and the world (civilisation and universe)," Tucker concludes that homes are "multileveled," and that the kind and number of levels of a home differ for each person (Tucker 1994, 182). Moreover, one's home is not a fixed structure, but rather, it changes over the course of a lifetime as experiences accumulate and conditions vary. "Most people spend their lives in search of home, at the gap between the natural home and the particular ideal home where they would be fully fulfilled," observes Tucker (Tucker 1994, 184). For one week each year, Jews ritually construct sukkot to bridge this gap to which Tucker refers and to dream of the ideal home in which their personal fulfillment resides.

In all four case studies presented here, three facets of the individual builders' lives are linked in a dynamic relation: their physical places in the world, the material expressions of their religious beliefs, and their notions of an ideal state of being—their dreams. The four ideal states of existence expressed in these cases, or what can be understood as notions of "home," are invested in a homeland, an ethnicity, a relationship with God, and the possibility of a just and equal society. These concepts of "home" align with idealism or virtuality as opposed to materialism or materiality; however, as meaningful as their contrast is their connection. All of these case studies demonstrate how the creative shaping and interpretation of our material worlds allow us to transcend them in the individual and the collective search for home.

Acknowledgments

This research would not have been possible without the wisdom, time, and spirit of Bakol Geller, Drori Yehoshua, Rabbi Chaim Halberstam and his family, Mayer Preger and his family, Joseph Piekarski, Rabbi Ephraim Piekarski, Dror Kahalani, and the residents of Shchunat Hatikva. They generously shared with me the meanings of the materials that define their practice. In the review and production of this work, Indiana University Press staff and outside readers provided valuable advice on both editorial and content matters for which I am most grateful. Finally, I am indebted to Jason Baird Jackson, who surpassed his duties as series editor in offering encouragement, opportunity, and mentorship from the beginning of this research project to its completion.

Notes

1. Although not emphasized in this analysis, another second key interpretation of this holiday's ritual architecture evokes the harvest history of the Israelites' agrarian past.

2. Although I pursue here the relevance of the notion of belonging in the context of Sukkot in the Jewish Diaspora and in Israel, folklorist Galit Hasan-Rokem (2012, 156) has importantly challenged the historical binary of periphery and center in a discussion of the Jewish Diaspora through her study of the sukkah. She analyzes the sukkah as a site where itinerancy and locality intersect, identifying the ritual structure as that which mediates between mobility and stability. Hasan-Rokem writes, "certain religious material practices have undermined the dominant idea of a concentric Jewish universe" (2012, 3), claiming the sukkah as a primary example.

3. According to statistics published in the March 2013 Israeli Authority on Population and Migration's "Report on Foreigners in Israel," available on the Israeli Government's website: http://www.piba.gov.il/PublicationAndTender/ForeignWorkersStat/Documents/foreign_stat_032013.pdf, accessed March 2013. In this report, as of February 28, 2013, the exact figures listed are 64,638 "infiltrators'" entered Israel; 55,195 "infiltrators" currently reside in Israel; 70,584 legal foreign workers in Israel; 14,549 illegal foreign workers in Israel.

4. All quotations attributed to named individuals from Hatikva come from my transcribed and translated fieldwork interviews.

5. Babylonian Talmud, Masechet Baba Meziah: Chapter 5, Page 71, A.

References

Bachelard, Gaston. 1994 (1958). *The Poetics of Space.* Boston: Beacon Press.

Fleischman, Danielle. 2011. "Occupy Wall Street Protesters Have a Sukkah." *Jewish Telegraphic Agency,* October 12. Accessed July 5, 2013. http://www.jta.org/2011/10/12/news-opinion/unitedstates/occupy-wall-street-protesters-have-a-sukkah.

Hasan-Rokem, Galit. 2012. "Material Mobility Versus Concentric Cosmology in the Sukkah." In *Things: Religion and the Question of Materiality,* edited by Dick Houtman and Brigit Meyer, 153–79. New York: Fordham University Press.

Jackson, Jason Baird. 2006. "Diaspora." In *The Greenwood Encyclopedia of World Folklore and Folklife,* edited by Thomas A. Green and William M. Clements, 18–22. Westport, CT: Greenwood Press.

Primiano, Leonard Norman. 1995. "Vernacular Religion and the Search for Method in Religious Folklife." *Western Folklore* 54:37–56.

Rubenstein, Jeffrey. 1994. "The Symbolism of the Sukkah." *Judaism: A Quarterly Journal of Jewish Life and Thought* 43:371–87.

Tucker, Aviezer. 1994. "In Search of Home." *Journal of Applied Philosophy* 11:181–87.

Turner, Kay. 1999. *Beautiful Necessity: The Art and Meaning of Women's Altars.* London: Thames & Hudson.

Turner, Victor. 1969. *The Ritual Process: Structure and Anti-Structure.* New Brunswick, NJ: Aldine Transactions.

GABRIELLE A. BERLINGER is Assistant Professor of American Studies and Folklore and the Babette S. and Bernard J. Tanenbaum Fellow in Jewish History and Culture at the University of North Carolina at Chapel Hill.

2 (Not) Going Public: Mediating Reception and Managing Visibility in Contemporary Scrapbook Performance

COMPILED BOOKS—COLLECTIONS OF autographs, quotations, trade cards, souvenirs, recipes, ballad texts, photographs, political commentary, and markers of child development—have been part of vernacular practice and popular literacy in the United States for centuries. We sort them roughly into genres based on subject matter, medium, contributors, even audience: names such as *autograph album, baby book, receipt book, community cookbook,* or *brag book,* for instance, call up specific content and contexts.[1] Yet the overarching term *scrapbook* can encompass them all, pointing simultaneously to these objects' incredible flexibility (in materials, structure, and meanings) and to their cultural invisibility (just idiosyncratic scraps of things and ideas, after all). As the twenty-first century approached, however, a certain kind of paper-based scrapbook was hard to ignore. The practice known as *scrapbooking* started gaining steam in the mid-1990s; by 2003, it had become a multibillion-dollar industry. According to one report, more than 98 percent of avowed scrapbookers at the time were women, many between the ages of thirty and fifty, and the majority were primary caregivers who worked at least part time outside the home (Primedia 2004).

These books were made to be seen—often literally constructed in groups, with some makers investing hours in a single page—and they were built to last. Crafter-archivist entrepreneurs created and marketed acid-free paper, archival-quality albums, photo-safe adhesives, and a range of decorative materials to adult women in the United States and around the world.[2] At the turn of this century, design

styles varied widely, from minimalist treatments to intricate collages; many practitioners studied the fundamentals of photography and graphic design, took writing classes to spark or to focus
their commentary, and incorporated a range of tangible and digital embellishments, from ribbons and beads to metals and vellum.
Though the practice persists in a range of forms and among varied
populations today, my discussion here focuses on scrapbooking
clearly connected to the burgeoning industry in the early 2000s.
Between 1999 and 2006, I explored the strategies and motivations of
women who self-identified as scrapbookers; much of my fieldwork
took place in the Midwest, where I was finishing a doctorate in folklore at Indiana University, but I also attended classes and conventions
in the mountain West and along the mid-Atlantic seaboard (see
Christensen 2009, 2011).

So it was that in 2003, Connecticut native Joan Potts Daniels—a
sixty-six-year-old mother of two and former florist and wholesale
book manager—sat down to show me her scrapbooks. Unlike nineteenth-century albums that have garnered acclaim for their intricate
handwork or inscrutable juxtapositions (Fig. 2.1), and unlike more
recent books that have celebrated a similar handmade, vintage, or
accretive aesthetic, Joan's pages seemed particularly straightforward
and efficient. Strap-bound and nearly twelve inches square, the
cardstock pages of the albums were overlaid with materials that
echoed the photographs but offered very little written contextualization. Joan stopped at one layout, whose photographs showed four
couples at a table spread with a yellow cloth and a potted centerpiece;
while one or two individuals in the photos acknowledge the camera
with a smile or a glance, for the most part they are eating, chatting,
listening, and reaching for or passing food. Four uncaptioned rec-
tangular snapshots filled the lower quadrants of the page, while a
fifth, cropped into a circle, was centered in the top third and banked
by small stickers that depicted multicolored corn; a smiling turkey; a
Pilgrim wearing a conical, flat-topped capotain and carrying a basket
of produce; and two matrons (one in buckskin and braids, the other
white-aproned) engaged in sympathetic conversation. The page's
only text—"Our Best Thanksgiving" and "1995," both spelled out
with colorful 1980s-era stickers—identifies the gathering as a late-
twentieth-century holiday meal (Fig. 2.2).

FIGURE 2.1.
One of more than two hundred pages in Anne Wagner's three- by five-inch *Libri Amicorum, or Memorials of Friendship* (1795–1834), this one contributed by her niece (Carl Pforzheimer Collection 2016). Complete album is available at http://digitalcollections.nypl.org/collections/anne-wagner-album-1795-1834#. (Photo credit: New York Public Library.)

But why, one might ask, would Joan preserve these images, which echo a thousand other snapshots in content and composition? Furthermore, why curate, contextualize, and display them in a vernacular book genre—especially if the resulting creation is apparently one-dimensional, embellished with mass-produced stickers but none of the other material layerings that have defined scrapbooks of yore? In what follows, I answer that Joan's page, and others like it, are in fact embedded in a rich matrix of narrative and intent; cultivating a seemingly unencumbered surface is rhetorically useful for controlling the reception of information and, indeed, for (re)defining the social contexts of private display and the social value of domestic experience.

The Self in Public: Genres of Personal Display

Multiple scholars have explored contexts in which individual lives are represented through material culture. Barbara Kirshenblatt-Gimblett has studied ensembles of household objects—from miniatures arranged in tableaux to rag-rug stair runners—that embody memory

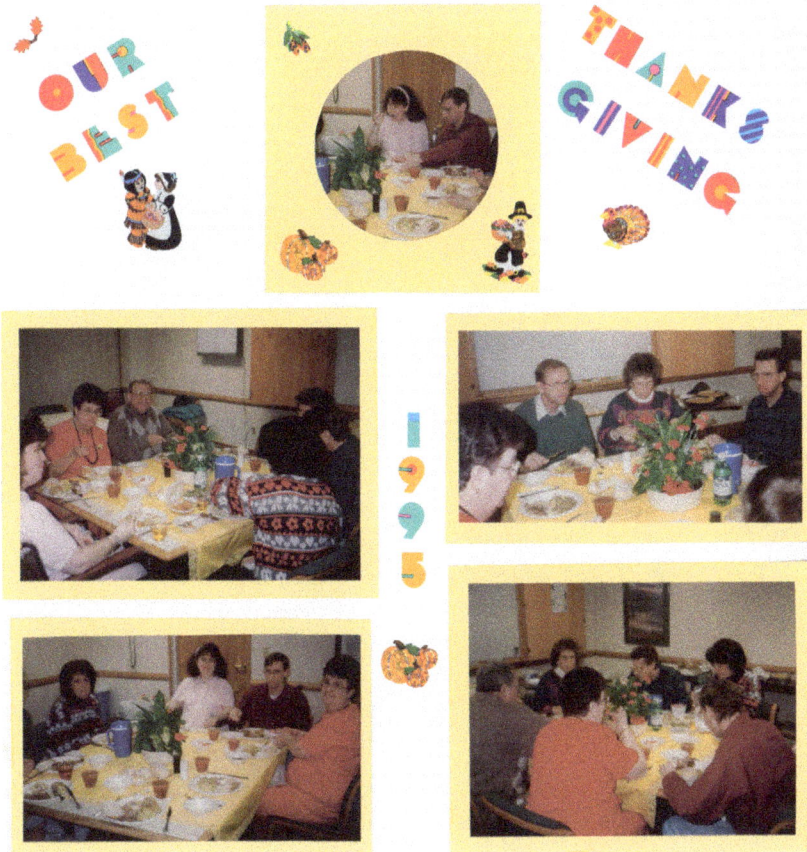

FIGURE 2.2.
Joan Potts Daniels (Bloomington, IN), "Our Best Thanksgiving," twelve- by
twelve-inch cardstock, ca. 2001.

and relationships and act as forms of life review (Kirshenblatt-
Gimblett 1989), and Henry Glassie has reflected on the meanings of
things in Ellen Cutler's Ballymenone kitchen (Glassie 1982).[3] Jennifer
González considers "curio-cabinets, boxes, drawers, shelves, niches,
[and] altars" as sites of "autotopography"—spaces in which "a compen-
dium of symbols and indices that represent personal links to other
times, locations and individuals" can be collected and displayed
(González 1993, 82). These examples of personal exhibition occur in
and transform relatively private spaces, while other everyday forms of
showcasing—yard decorations, state fair entries, door collages—are

oriented to broader audiences. Bumper sticker collections reach more viewers than do staircase galleries.

Displays lodged between the covers of an album seem especially private—particularly if their contents are minimally labeled. Not surprisingly, scrapbooks are sometimes assumed to be retrospective archives, a way to contain and muse upon the selected flotsam of the individual life.[4] Certainly, these books (and the processes of making them) offer their creators opportunities for personal reflection. But approached ethnographically, scrapbooks can also be experienced as exhibits that makers use to position themselves actively within symbolic discourses and regimes of value (cf. Appadurai 1986; Kirshenblatt-Gimblett 1995), as well as in relation to live audiences, both familiar and unknown. Here, I draw on theories of genre, textuality, and performance to argue that scrapbooks and scrapbooking are modes of display that monitor levels of disclosure as they negotiate the benefits and gendered risks of "going public."

I am interested in publicness because it matters in social life. The terms *private* and *public* form an ideological complex that is used to call up (and differentially value) distinctions between spaces, motivations, types of interactions, kinds of work, and spheres of influence. However, linguistic anthropologist Susan Gal has argued that the public/private divide is, like so many dichotomies, a fractal distinction rather than a line drawn in the sand; not only do the two domains blend and intertwine, but they do so recursively. That is, a distinction made at one level of specificity can be mapped onto another (Gal 2002). (Some private space, for instance, is more public than others, as in the parlor/kitchen distinction documented by scholars of vernacular architecture [e.g., Glassie 2000; Pocius 1991].) These degrees of relative public- and privateness are constructed and signaled through the ways that people talk and act (Gal 2002). Moreover, discursive calibrations of public and private have social consequences: historically, "public" expressions—those intended for broad distribution and focused on abstract or universal themes—have been associated with political, economic, and intellectual power.

Thus, I want to know where and when ideas and behaviors become visible, and also more about *what* is deemed worthy to be seen (or indeed, recognized at all). In this chapter, I argue that contemporary scrapbook makers participate in a politics of genre by redefining what is appropriate to broadcast widely.[5] Scrapbook texts and practices that have emerged since the 1980s disrupt genred expectations regarding where and how the self-disclosure typical of diaries, letters, and other heart-to-hearts should be performed; scrapbook makers I have worked with draw on features of several genres (and have cultivated new contexts of display) in order to access the social influence associated with publicness and use it to claim value for *carework* and *kinwork* (Abel 2000; Di Leonardo 1987; Meyer 2000): for the labor, skill, and creativity that go into tending people and social networks. At the same time, participants critique what it means to go public. In effect, they create displays that are public enough to expand personal visibility, yet still practice the supportive intimacy idealized as typical in "private" contexts (Katriel and Philipsen 1981)— those in which interlocutors are better known and more concretely circumscribed.

When considering how makers respond to the social hierarchies structured by public/private dichotomies, the materially composite nature of scrapbooks is as important as the ways the books are used and shared. Albums stumbled upon in archives, at thrift stores, or on a neighbor's coffee table might appear to be unconsciously forthright or, at the other extreme, superficially whitewashed. Scrapbookers may in fact cultivate intentional invisibility, fabricating their books in ways that keep intimate information private by hiding it in plain view. For instance, some use the scrapbook's literally layered form to conceal information inside or beneath other objects. But they may also cultivate hypervisibility, favoring material forms and social groupings so apparently transparent that they don't merit deeper investigation. Troublesome topics are not necessarily buried or ignored, however; I've found that in oral performance—in the narration of objects—conscious elisions or cursory treatments are generally reinserted or complicated. Let us examine, then, how contemporary scrapbookers play with genre conventions, material substance, physical spaces, and talk as they seek to control meanings, mitigate censure, and redefine the terms of public topics and behavior.

Studying Surfaces: Unpacking the Allographic Text

At a purely visual level, many scrapbooks are replete with unexplained smiles and cheery mass-produced papers; consequently, it is tempting to dismiss vernacular albums as the slapdash (if harmlessly sentimental) arrangement of commodities that include everything from Polaroid instant photos to themed embellishments, such as holiday stickers. In recent decades, "amateur" photographs have been noticed by some aesthetic connoisseurs precisely because of their perceived *in*significance, a valuation that stems from reading these snapshots as unthinking reflections of technological and social convention. In 1944, Willard Morgan, director of the Museum of Modern Art's Photography Center, characterized snapshots as "folk art" because he saw them as "honest," "unselfconscious," and "spontaneous, almost effortless" cultural productions (Chalfen 1987, 73). Sometimes described as "home mode" photos, these visual objects have been perceived as naïve records rather than deliberate artistic efforts (Chalfen 1987, 135; Musello 1980). In fact, critic Janet Malcolm has asserted that it is the home snapshot's "ineptitudes and infelicities," "haphazardness, capriciousness, and incoherence" that have permitted (professional) postmodern artists to appropriate these conventions ironically and "court formlessness, rawness, clutter" as an aesthetic rather than producing such disorder by accident (Malcolm 1976).

Vernacular compilations of snapshots have also been too quickly considered by cultural connoisseurs. The album—a "secondary" genre that corrals and contextualizes other discrete forms (Bakhtin 1986)—predates photographic technologies; book genres have been filled with quotes, inscriptions, mementoes, and illustrations for a very long time (Fig. 2.3). But while commonplace books—collections of quotes and ideas—are linked in popular discourse to John Locke, Enlightenment goals, and the study of rhetoric, scrapbooks have less lofty connotations. Explaining that I've studied recent scrapbook enthusiasms in the United States has often elicited a polite "Oh," or a surprised, "Really?" from casual academic peers. Long associated with women, children, and the construction of visual and interpersonal concord, scrapbooks and their creation may be perceived as an essentially "safe"—and thus largely impotent—sentimental and backward-looking endeavor. These linkages are not entirely unmotivated; it is true, for instance, that some of the more than 2500 specialty stores that sprang up around scrapbooking in the early 2000s sported monikers such as "Forget Me Not," "Recollections," "The Paper Attic," "Memory Lane," or "The Paper Daisy"—names that evoke a wistful preoccupation with a brittle past or the fading joys of a simple

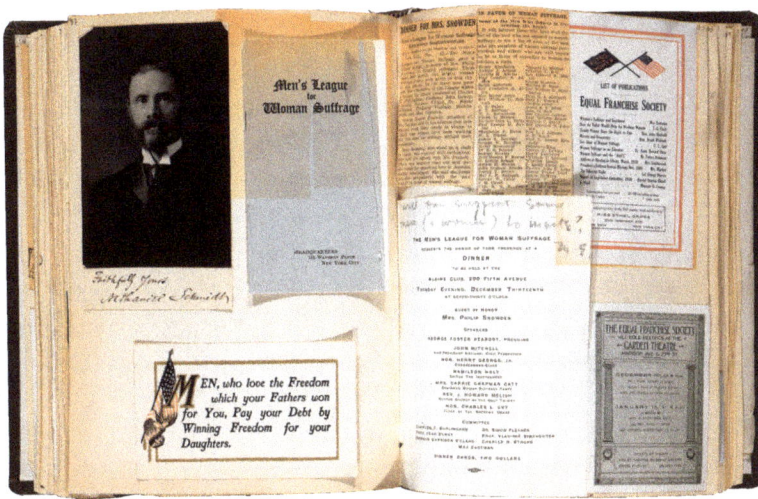

FIGURE 2.3.
Pages from the scrapbooks of Elizabeth Smith Miller and Anne Fitzhugh Miller bring together photos, signatures, programs, brochures, and newspaper reports that document male allies in the fight for women's right to vote, ca. 1910. "Men's League for Woman Suffrage, Miller Scrapbook," National American Woman Suffrage Association Collection, Library of Congress, LC-DIG-ppmsca-02966.

childhood. These shops fostered nostalgia in the common sense of that word: appellations like "Memory Lane" suggest a desire for a selective, idealized, and static—even "embalmed"—past, or an acute awareness of an insufficient present and a "(painful) longing to return" to a better time (Jones 1995; McDermott 2002, 390).[6]

Thus, if vernacular snapshots and scrapbooks can be characterized as innocuous pastimes lacking intention or depth, they can also be interpreted in a more sinister light, as reflections of oppression or as tools of denial. One might argue that, like other hobbies associated with (white) middle-aged, middle-class mothers—crazy quilts, romance novels, soap operas—scrapbooking is a consumption-based compensatory escape for women bored by privilege or trapped by family demands and the social conventions of patriarchy (cf. Ang 1985; Gelber 1999; Radway 1984 [1991]).[7] Because the acts of curation and contextualization that define the practice enable editing, elision, and framing, scrapbooks can also be viewed as outright mis-representations. Such is the case in one satirical commentary—an article in *The Onion* called "Local Woman's Life Looks Bearable in Scrapbook"—that imagined a fifty-eight-year-old homemaker named Mrs. Hemmer showing off a book that "project[s] an image of a functional family bound by unconditional love and total fulfillment. By layering carefully chosen photos with brightly colored paper, elaborately patterned borders, and whimsical stickers, Hemmer has successfully concealed a lifetime of anguish, scorn, and contempt" ("Local Woman's Life" 2004). Her creation, the spoof went on to suggest, leaves out the squabbles, alienation, and distraction that are the stuff of actual human intimacies.

These critiques offer useful correctives to the celebratory mode that dominates scrapbook industry discourse, and they help to explain some social meanings of the current practice. But they also generate a critical context in which scrapbooks and scrapbooking are understood to be both simply transparent (their meanings and motivations obvious to the casual observer) and simply opaque (with slick and impermeable surfaces that conceal darker realities). I believe that approaching these books as situated performances allows a more nuanced set of conclusions.

This, for instance, is how my interaction with Joan unfolded on that November day in 2003 when we looked at her representation of Thanksgiving 1995.[8] As she turned to the layout, Joan recalled, "This one—this is when Katie was born." Joan's only grandchild at the time of our interview had come into the world several months early, weighed just a few pounds, and stayed in the hospital for eighty-seven days. Indeed, on closer inspection, the page offers hints of an industrial setting: a large whiteboard on one wall, a vinyl accordion room divider, a tablecloth that doesn't quite fit the dimensions of the laminate table it dresses. Joan continued,

> And this is in the hospital. She was born three days before Thanksgiving.
> And we didn't know if she was gonna make it.
> So I put "Our Best Thanksgiving" [as the title] because
>   we ordered Turkey dinner from Krogers, and picked it up,
>   and the hospital gave us a room of our own,
>   so this was, the whole family—
> So I says "Our best Thanksgiving" because Katie *made* it. (Christensen 2003)[9]

Katie's mother, Susan McDonald Daniels—herself a scrapbooker raised in Nebraska and later employed in the papercrafting industry—documented the entire hospital stay in a different album but years later had yet to complete her journaling for these pages. She had not, she told me, found the words to accurately convey the emotions that the photos evoke; a two- by three-inch space seemed inadequate to explain what it felt like to plead for the life of a premature infant, a child conceived after much anxiety. Susan's pages reanimate the fears that accompany parenthood and also confirm her sense that Katie is a gift; thus, they are contemplated in bittersweet relation to the present, rather than indulged in as an escape to happier times.

For Joan, including only a title and date was a choice of a different kind. Assuming that most viewers of her album wouldn't be familiar with her granddaughter's life, she deliberately selected evaluative words that would prompt inquiries. "See," she told me, "they wouldn't know [the story behind the photo.

> The title] just says "Our *Best* Thanksgiving."
> "Well, why?"
> So then I, *explain.* (Christensen 2003)

Structuring the page and its discourse this way, she said, "gives me a chance to express what the page is about, without actually writing a whole story on it." Joan set out to create a literal conversation piece that was fully realized only during face-to-face encounters.

Joan's work demonstrates that albums that include "minimally explicit" messages (Tannen 1985, 128, 130)—those that are content to hint at meaning through a few facts or well-worn phrases, or that eschew explicit notations entirely—do not necessarily represent lack of time, care, or ability.[10] These kinds of scrapbooks depend on, and are enriched by, verbal performances. There are, of course, scrapbooks that spell out every name, date, and story in writing; makers have rendered these books detachable from their original contexts in a way that is maximally fixed. However, entextualized discourse can encourage emergent performance if it is packaged in a way that requires further elaboration.[11] "The need for commentary is enhanced," observes anthropologist Karin Barber, "when the formulation is allusive, opaque, truncated, or otherwise obscure" (Barber 2005, 269–70).

As an example, Barber cites Kwesi Yankah's 1994 account of a bone suspended from string at a proverb-custodian's home in Ghana, an object that posed a puzzle to people who encountered it. Why was that bone hanging there? Visitors who asked learned that the bone represented a newly minted proverb—and they also heard the story behind the proverb's invention. This object, Barber concludes, became understandable when "bathed in a sea of contextual and historical detail, which is not encoded *within* the object or the proverb but is transmitted in another genre—the personal narrative—that runs alongside them" (Barber 2005, 269).[12] Thus embedded in dialogic routines, the bone is an object that requires both author and audience to play active roles in constructing meaning.

Scrapbook makers craft similar "involvement-centered" texts (Tannen 1985, 125) using several strategies. Some seed pages with catchy narrative "kernels" (Kalčik 1975)—such as the alliterative title "Maker of Much Mischief"—that index and ostensibly call forth a longer story. Others draw on witty or familiar phrases (e.g., headlining a page about an independent-minded daughter "Pure Spice" in order to revise a well-known nursery rhyme about what little girls are made of; Fig. 2.4). Some pose explicit questions (in page titles, or in speech balloons layered onto images) that makers answer as they peruse pages with their audiences. Joan, however, was unusually explicit in discussing the way she used *elision* as a form of strategic engagement.

While explaining how she chose photos for her pages, for instance, Joan told me that quantity wasn't important: "Like the one with Katie," she said, referring to a single twelve- by twelve-inch page dotted with Band-Aid images and minor exclamations of pain (Fig. 2.5). "There's only two pictures on that page, but it tells a whole story." In fact, the page itself—which features two photos of a small girl with a skinned nose and a bruised cheek—does *not* tell a whole story (at least, not on its own). The cartoonish paper behind the snapshots does suggest that the narrated event won't be particularly sinister. However, the mere existence of these photos is enough to prompt a query. In his surveys of snapshot photography, Richard Chalfen found that parents have been unlikely to photograph injury, unless (like missing teeth) it connotes developmental progress or (like surgery) represents a milestone (Chalfen 1987, 30). A viewer familiar with these conventions would be curious: Why did these

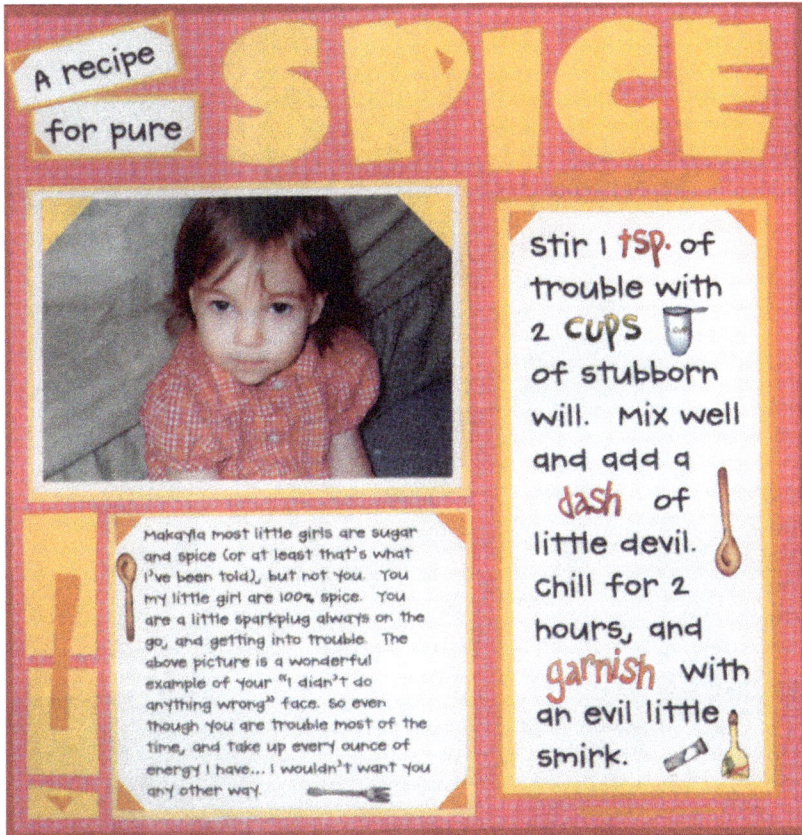

FIGURE 2.4.
Michelle Tardie, "A Recipe for Pure Spice," published in *Ivy Cottage Scrapbook Magazine*, February/March 2003, 82. This layout references a well-known nursery rhyme but also invents a "recipe" that calls to mind the metaphorical directions for "preserving a husband" often found in compiled cookbooks. The explanatory text (lovingly) notes that the child is "trouble most of the time" and "take[s] up almost every ounce of energy I have."

wounds receive special recognition?[13] As Kathleen Stewart has observed, "gaps in the meaning of signs" create "a place for story" (Stewart 1996, 3).

After briefly narrating the story behind Katie's injuries, Joan turned to another layout—again epitomized by narrative brevity—whose central image is of the child moving pieces on a chessboard (Fig. 2.6). As in much of Joan's work, the content of the photo was underscored by icons (black-and-white checked background paper;

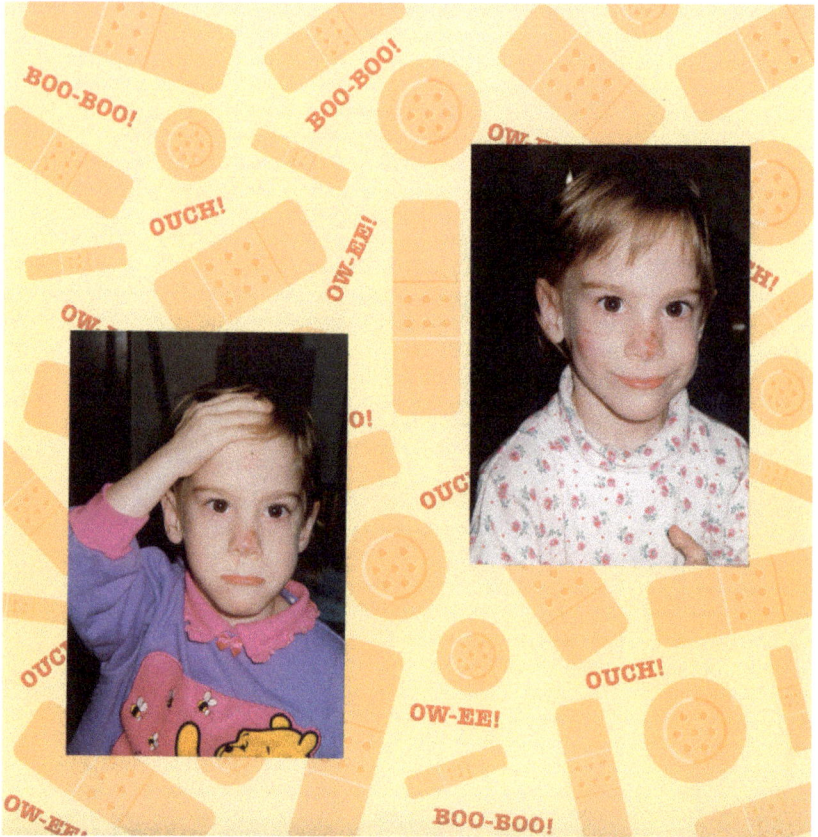

FIGURE 2.5.
Joan Potts Daniels (Bloomington, IN), "Ouch!" twelve- by twelve-inch patterned paper, ca. 2001.

silhouettes of a pawn, a rook, and a knight) and alluded to in the handcut title "Checkmate." As we paused here, Joan continued,

> But it, like, *here,* [an onlooker would ask],
> > "Is she really playing chess?"
> And I says,
> > "Yes, here—"
> She knew, at that age, she knew the pieces—she knew the pawns from the rooks and everything. (Christensen 2003)

Notice that Joan not only anticipates viewer questions, but in her talk with me she actually enacts this exchange as she speaks, using unattributed direct discourse to play the participatory part she

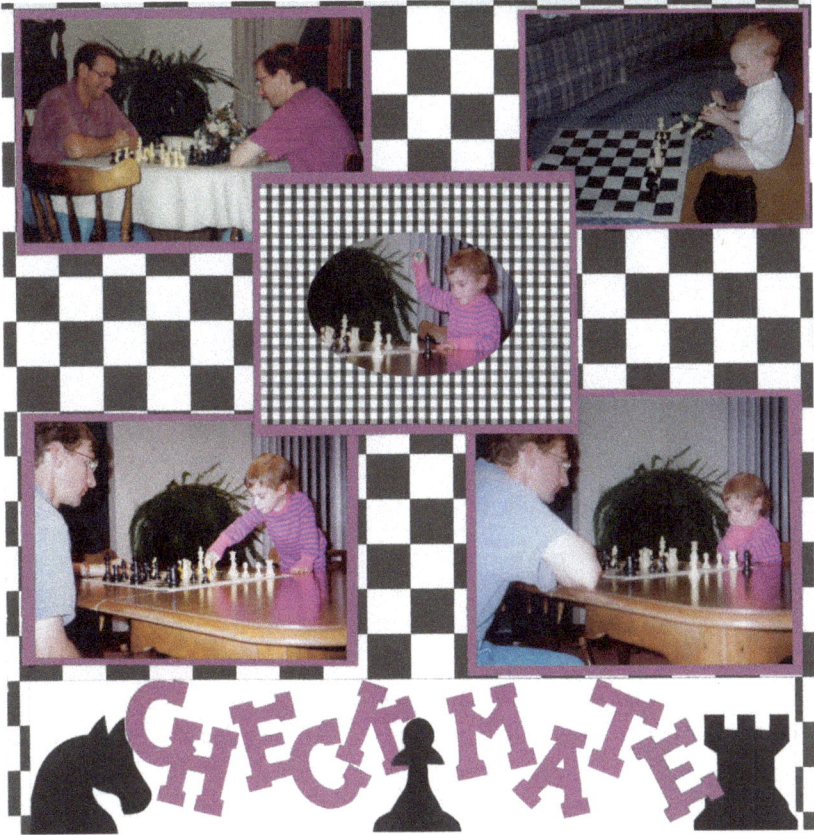

FIGURE 2.6.
Joan Potts Daniels (Bloomington, IN), "Checkmate," twelve- by twelve-inch cardstock, ca. 2002.

expects of her audience. Once interest has been established, Joan doesn't necessarily wait for additional questions before explaining her work. Immediately after quoting her response to this potential viewer, she continued, "But—so if they ask questions like that, then, you know—this, this is a page, I have to tell you about this one."

Like novels or diaries, scrapbooks are in one sense *autographic* texts that can stand alone as meaningful messages: they are books, after all. Yet like a script or musical score, many scrapbooks are also *allographic* (Goodman 1968), crafted with the expectation of repeated but always unique bodily enactments before an audience. Furthermore, involvement-centered books like Joan's carry what sociolinguist Deborah Tannen calls "a metamessage of rapport" (Tannen 1985, 125). Involvement-centered communication generally involves a more private public—a one-on-one encounter with known interlocutors; thus, Joan may "leave out" information (favoring the allographic) because she prefers the pleasure of filling in the gaps orally before an engaged, copresent audience. But because an allographic text guards its secrets well, it also usefully anticipates (and deflects) the gaze of strangers. Before discussing specific strategies for mediating disclosure in this way, let me establish the contexts in which strangers are invited to observe these pages and suggest how contemporary scrapbookers have sought to mitigate the attendant consequences.

Claiming Value: Genre and Carework

In *A Room of One's Own,* Virginia Woolf gave this instruction to women: "Above all, you must illumine your own soul with its profundities and its shallows, and its vanities and its generosities, and say what your beauty means to you or your plainness, and what is your relation to the ever-changing and turning world of gloves and shoes and stuffs swaying up and down among the faint scents that come through chemists' bottles down arcades of dress material over a floor of pseudomarble" (Woolf 1929 [1989], 90). One might envision Woolf's addressee sitting down to reveal herself in a diary, a letter, or even an autobiography (genres that envision progressively wider audiences). Or perhaps she would choose a photo album, a genre predicated on display: one student of visual communication has suggested that "viewing events are perhaps the culmination, or even the very point," of vernacular photography (Musello 1980, 33). Taking a stance on a public stage is a way of claiming value, a path to reflection, recognition, and expanded networks.

Yet exhortations to "illumine . . . the profundities and shallows" of individual lives hardly erase expectations regarding what can (or should be) publicly disclosed. After the development of the Kodak Brownie in the late nineteenth century, only a small portion of everyday life was regularly captured by amateur photographers, and only a limited range of snapshots has been considered relevant to the public gaze. Scholars in the 1980s found that photograph albums overwhelmingly documented a small circle of people engaged in the pleasant, the public, and the unusual (Chalfen 1987; Musello 1980). Personal experience genres that have been recognized as verbal art—tall tales, yarns, memorates—also emphasize the extraordinary (Bauman 2004, 112; Dégh and Vazsonyi 1974; Honko 1964; Wilson 1995).

Scrapbooks, too, have tended to privilege unusual or "special" events, offering evidence of personal participation in activities already deemed worthy of broad acclaim (Katriel and Farrell 1991, 15). Many contemporary scrapbook subgenres continue in this tradition: while the self-conscious brag book is the most overtly celebratory form, heritage, holiday, child-centered, retirement, and milestone (baby, wedding, bat mitzvah) albums also incorporate praise as their dominant mode.[14] Scrapbooks by or about men—or at least those preserved in archives—tend to be even more public, focusing on paid, professional, or competitive pursuits (including athletics, hunting, debate, theater); they are, in effect, "paper-and-gum monuments" (Ott 2006, 31; cf. Katriel and Farrell 1991, 3–4).

Then, too, Sidonie Smith has suggested that autobiography itself is an "androcentric genre" that "demands the public story of the public life"—an expectation that renders much domestic experience unrepresentable (Smith 1987, 52). (Note that Woolf specifically asked women to articulate their relationships to public displays of consumption.) When the vagaries of domestic experience have been shaped into durable form for public audiences, as in novels or poetry, they have often been dismissed as sentimental or embarrassing. Nathaniel Hawthorne's famous condemnation of "scribbling women," for instance, stemmed in part from his distaste for "indecorous exposures of the personal, the familial, and the bodily" (Wallace 1990, 203). Though Hawthorne's own work often explored the ethereal, the mutable, or the moral, he nevertheless criticized women writers for "parading before the world personal problems, domestic squabbles, and medical curiosities that ought to be suppressed" (Wallace 1990, 209).

Even book genres intended to spotlight the intimately personal, such as the details of a baby's growth and development, can reinforce the assumption that the work of primary caregivers should remain invisible. My own commercially produced baby book from the early 1970s is filled with preprinted sentences that begin "My name is _____," and "Today I _____"—phrases that encouraged my mother to speak *about* me in *my* voice, thereby muting her own. These phrases also tend to give the child credit for tasks actually completed by the parent (see Ochs 1992). My book congratulates me for "going to the doctor" or "eating my first steak," with no suggestion of how Pamela Little Christensen juggled her time and our family's budget to make those activities possible. How is it, then, that experience culturally designated as "private"—"the personal, the familial, and the bodily"—might be granted the acclaim long afforded to activities designated as "public?"

One option is to play to the functional strengths of the scrapbook as a genre. Precisely because scrapbooks have historically been an epideictic form—one filled with things that evoke praise and recognition (Katriel and Farrell 1991)—these books can convey noteworthiness on content that might not otherwise be perceived as such. By deliberately documenting the "backstage" work that makes special events distinctive, scrapbook makers revalue daily life by recontextualizing it, making it, in effect, "reportable" (see Appadurai 1986; Clifford 1994, 265; Hymes 1975). In this sense, contemporary scrapbooks do not just *save* and archive evidence of public acclaim; rather, some are created as tokens for, and vehicles of, this kind of acknowledgment with regard to work and interests that are typically unseen and unremarked.

At the turn of the millennium (and as a precursor to "mommy blogs"), scrapbook enthusiasts increasingly documented the mundane aspects of everyday life—the kind of social and subsistence tasks that constitute carework (Abel 2000; Meyer 2000). In 2004, for instance, Deirdre Paulsen, an early (1980s) proponent of archival practices among lay family historians, shared with me a small scrapbook that her daughter-in-law, Sarah Condon Paulsen, had given her. The title page identified the book's subject and erstwhile "speaker"— "A week in the life of Tessa, November 2003 (almost two years old)." An inscription flanked by hand-drawn holly leaves indicated the book's status as a Christmas gift, and the photos—which are most frequently of Tessa or Tessa with her father—suggest that they were taken by Sarah, the other primary actor in the book.

At the time, Sarah was a thirty-one-year-old mother of two living in Muncie, Indiana. Originally from Spokane, Washington, she earned her BS in biology teaching and taught middle school for four years, until Tessa was born and the family moved to begin medical school in Minnesota. During her husband's year as a resident intern in Indiana, Sarah was the primary caregiver for their daughter and

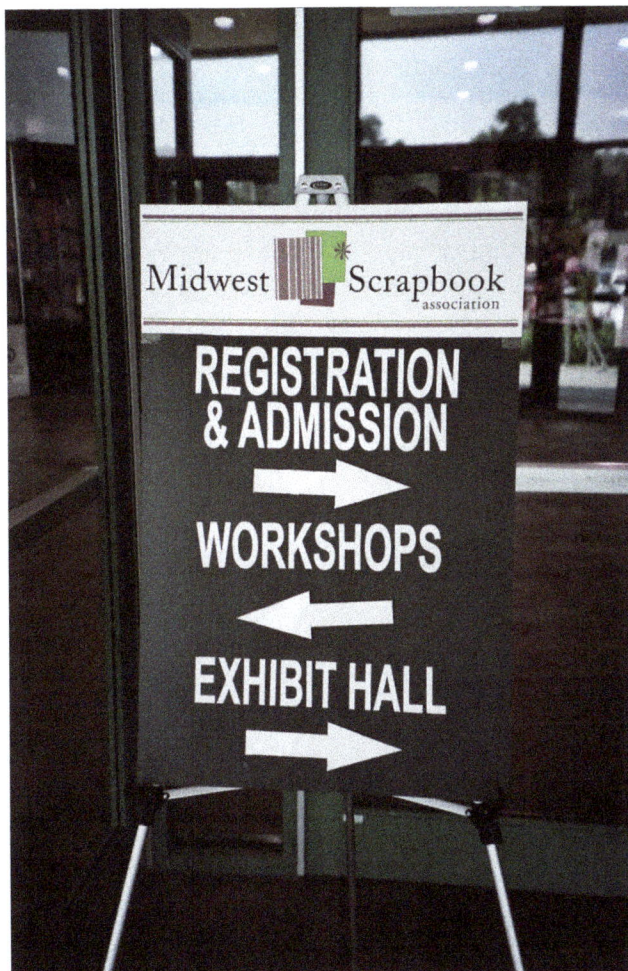

FIGURE 2.7.
Those who attended the Midwest Scrapbook Convention in September 2006 could participate in dozens of classes, view contest entries, visit vendor booths, and stay to create and socialize during evening "crop" events. Indiana State Fairgrounds, September 2006. (Photo by author.)

infant son. In late 2003, Sarah decided to make a small book about a week in the life of Tessa, then her only child, and present it to her parents and her husband's family; the idea had been suggested by a community child-parent class she attended in Minnesota, but she modeled the album on one that her sister had made back in Washington State.[15]

The book, which consists of eight 11- by 8.5-inch black cardstock pages, is handbound with wire-edged ribbon, using bookbinding skills Sarah learned in college. The A. A. Milne quote displayed on the cover—matted in purple on pale yellow paper splashed with bright watercolor flowers—proclaims the point of view that pervades the album: "When I was one, I had just begun." The computer-printed columns that describe each day's events are also resolutely first-person: "Today was another hard morning. I needed Mommy to rock me after I ate breakfast." . . . "When Mommy was checking out, Daddy and I went to the optical center and he tried on glasses." However, in spite of adopting Tessa's voice and appearing in only a few photographs, Sarah permeates this album. She is present, for instance, in the evaluations attributed to Tessa and sprinkled throughout the text (emphasis mine):

> "*I had a hard time waking up* this morning. Mommy had to rock me until I was ready to get dressed for church. That made us late. . . ."
>
> [at the dinner table] "*I decided I'm much too big* to sit in my booster seat!"
>
> "*After playgroup I was kind of grumpy* so Mommy put me to bed at 12:00."
>
> [at a neighbor's house] "*I was pretty tired* and didn't want to take turns on the swing."
>
> [at a restaurant, with friends] "*I enjoyed* standing on the chairs and *playing* with Daddy's water and lots of napkins. I kind of made a mess."

Processing daily tasks by fixing them in this material interpretation is an endeavor that reveals a complex of overlapping goals, labors, and relationships. For instance, if Sarah's entextualization of her child's life as a gift to her in-laws can be considered an extension of kinwork, and if the process also helps to manage daily carework, she found ways to record her own life, inclinations, and strategies in this small book as well. As each day in Tessa's life unfolds, we learn that Sarah goes to aerobics three times a week (except once when Tessa doesn't sleep well and so is "pretty grumpy," a situation that terminates the exercise session), reads/showers/works at the computer while Tessa plays or watches a video, visits the library, participates in a structured play-group, babysits a friend's child while the friend takes night classes, does some sewing, spends two evenings doing church work, takes Tessa to a playdate and then to visit her father (who is staying overnight at the hospital as part of a research study), attends a community parent/child class, goes shopping twice (for groceries and as a family outing), eats at a restaurant with a friend's family, and has an evening out with her husband at a dinner group called Gourmet Club. The album details baths, naptimes, food preparation and consumption, bedtime routines, and play, but it also demonstrates how Sarah manages her active schedule: she trades babysitting (when her friend's husband gets home from work, Sarah leaves her own daughter with him at their home); her husband participates actively in reading aloud, bathing, housework, and childplay when he's home; members of the Gourmet Club take turns watching all the children while their parents convene somewhere else. Sarah's/Tessa's book is thus a complicated entextualization of everyday experience, an object that makes visible multiple subjectivities. Yet it remains an oblique way for Sarah to reflect on and reveal her choices, struggles, and interests, while maintaining the other-oriented stance often socially valorized as evidence of The Good Mother.

"Going public" also encourages room for reflection and critique, however, and as the scrapbook industry developed, so did admonitions to speak more directly about personal and domestic experience. Scrapbooking had been a visible commercial sector for just a handful of years when magazines began to tackle the issue of "missing moms." In autumn 1998, the popular *Creating Keepsakes* published a feature called "Putting Yourself Back in the Pictures." Author Tavel Cowan-Bell shared the stories of a scrapbook storeowner and an instructor who realized, after working on books for several years, that they rarely appeared in their own albums. Cowan-Bell suggested that readers ask strangers or family members to step in, use an automatic timer to carve out dedicated time, or hire a professional photographer—and above all, "fac[e] the reasons you don't want to be in pictures in the first place," whether the reluctance stemmed from dissatisfaction with personal appearance, a preference for focusing on family, or the pleasure of being the photographer (Cowan-Bell 1998, 44).

Despite this early encouragement, as 2001 started, *Memory Makers* magazine still felt it necessary to advise: "Include yourself in your albums." The accompanying illustration on the January/February cover was a six- by six-inch spiral-bound "Book About Me" that featured a photo of a smiling woman. Her arm placement suggested that the photo was a spontaneous self-portrait (the now ubiquitous "selfie"), and the feature's lead-in confirmed this interpretation: "Your albums are filled with your kids and other family. It's time to turn the camera around and learn the benefits of creating pages about yourself." The implication was that mothers must take matters of public visibility and recognition into their own hands. The article began by summarizing a scrapbook class in Fort Collins, Colorado, that encouraged women to "reconnect with their inner selves—those parts of them that had been lost or forgotten after years with spouses, children, and careers" (Rueger 2001, 58). Sidebars and images throughout the essay suggested ways to ensure that personal photographs existed at all, and they offered hints for documenting individual reflections and material contexts. At times, author Lydia Rueger mobilized expectations of altruism to justify these self-centered creations. Quoting one "All About Me" advocate, she wrote, "Even if women aren't comfortable including themselves, Cheri [O'Donnell of Orange, California] urges them to think of their families. 'No matter what you think of yourself now, you'll appreciate having pictures of yourself years later, and your children will appreciate having them,' she says" (Rueger 2001, 61).

However, desires to produce more explicitly self-revelatory albums fly in the face of private/public boundaries encapsulated in conventional understandings of diary, journal, and scrapbook. Tamar Katriel and Thomas Farrell, for instance, contrast diaries—"introspective, opening interpretive verbal accounts of one's inner life"—with "items selected for inclusion in a scrapbook," which instead "provide a look at the self from the outside, as it were" (Katriel and Farrell 1991, 9). Thus, many of the avid scrapbookers I consulted between 1999 and 2006 found this kind of disclosure challenging. Although Michelle Allred and Amy Tippet, of Burlington, North Carolina, began their foray into scrapbooking by sorting through ancestral photos, by 2004 these neighbors had started to work on "All About Me" albums.[16] But they found it hard to "put themselves on a page":

Michelle: We are *struggling*, we are *struggling!*

   Amy: Well, because it's hard, because it's like a, it's like your own personal journal. And [you don't want someone to read that,

Michelle: [It's a big therapy group session, you know! You don't even know if you want to *put* that out there—"Are we gonna do this, and then *hide* it somewhere?"

Danille: [laughs]

Amy: I know, that's what I *said

Michelle: I mean, do we have to put this on the coffee [table]

Danille: Ah-huh.

Michelle: We, we're tryin to decide, which way we even want to *go* with it

Amy: Mmhmm

Michelle: because, it could get so, so personal.

Amy: I've got *one* page, done in mine. And, I can't—I mean, I know what I *want* to do, but I can't go any further, because, you open—you are opening yourself *up*, to—everyone can see, the inside

Michelle: And you don't want it to be this stupid little superficial thing, either;

Amy: Right

Michelle: Why waste your time on superficial? You know, but—you just have to *decide,* how far you want to go, with it, you know?

I asked for a little more elaboration on why the process was so challenging, and Michelle reiterated the genre trouble they were encountering:

Michelle: It's like taking your diary, and adding a picture to it, and *putting* it on the coffee table.

Danille: Oh, because it has a different audience—the diary is definitely, *just* for you

Amy: Right.

Michelle: That's locked up, in a drawer somewhere, you know,

Amy: Hid under the *mattress. . . . (Christensen 2004a)

Despite the difficulties of dramatizing the everyday and publicizing the private, I did encounter many scrapbooking practitioners who were willing to negotiate these complexities because, as Michelle said, "You want to be *known.* You don't want to be that picture, in that box, that nobody knows who you are. . . ."

Turning esoteric information outward (see Katriel and Farrell 1991, 9) involves textual shifts in diary discourse, much like the narrative elaboration that storytellers add when performing on stage or before a tape recorder (Bauman 1986, 105) or that a memoir writer might use when composing "for myself and strangers" (Bloom 1996). Indeed, some scrapbookers "preminisce," envisioning their albums as utterances that will be reanimated by others, in front of a future or unknown audience. "You've got to write as if you won't be there," one scrapbook maker told me. "Because you won't." Aiming both to address an imagined future public and also narrow intertextual gaps that might emerge between authorial intent and relayed performance, these practitioners craft "message-focused" albums: in other words, they spell out referential content, making its meaning less attributed to, and less reliant on, the interventions of those who engage it (Bauman 2004; Briggs and Bauman 1992; Tannen 1985). Those concerned with losing meaning after their creations are released into the world tend to include complex and detailed narratives, as well as directive captions that avoid deictic markers (e.g., personal and relative pronouns); some favor declarative sentences rendered in the third person. These albums are made coffee-table-ready by means of carefully shaped material themes and maximally stable written texts.[17]

In pursuit of increased recognition and visibility, scrapbookers have revised more than the *content* of their books; they've also sought out new audiences. Katriel and Farrell note that "scrapbooks—like diaries—are commemorative texts . . . designed for self-consumption"; they become "public territory" only if one can find others who "would care enough to delve into them" (Katriel and Farrell 1991, 8–9). Documenting special events—group reunions or holiday gatherings—may be one way to "politely exten[d] the audience for a snapshot" (Chalfen 1987, 82) by incorporating more people and by building in recurring opportunities to view the photos in which those people figure. Gifting scrapbooks is also a way to disseminate one's ideas and invite commendation from extended publics; as an other-oriented gesture, it's also a conventionally appropriate way for women, especially, to network (see Ulrich 2001; Weiner 1992). But scrapbook makers have also created novel territories for the display and evaluation of their exhibits.

Public Venues: Cultivating Recognition

Especially during the height of paper-based scrapbooking, album creation often took place at events called crops, "focused gatherings" (Goffman 1961) that participants likened to quilting bees or knitting circles. The crops I attended usually involved food, high spirits, lots of conversation, and very few children. Participants at any given time might share a number of ties—family, neighborhood, professional, religious—but for some, their social networks intersected at just this one point, around a shared practice (Noyes 1995). Although some crops took place in individual homes, the material constraints of scrapbook construction—the need for large workspaces, shared tools, specific supplies, time uninterrupted by family needs—quickly generated extradomestic venues in which relative strangers encountered each other and their albums. Scrapbooking-in-public has taken several forms, including commercially sponsored small-group events, come-and-go conventions, extended retreats, and large-scale instructional "camps" that focus on self-reflexive celebration. Thus, the creation

FIGURE 2.8.
A local scrapbook store in tiny Colchester, Illinois, July 2011. (Photo by author.)

FIGURE 2.9.
Visitors to the Indiana State Fair peruse the scrapbook page entries displayed at the Home and Family Arts building in August 2006. (Photo by author.)

and exhibition of scrapbooks became a dramatically more public endeavor. No longer displayed primarily at family gatherings or on the sideboards and coffee tables of "company" rooms, albums found new contexts for evaluation in print and online publications and in community and commercial space dedicated to group work (Figs. 2.7, 2.8, 2.9). Unless "your kids are showing them to their friends," Idaho-born Troba Nelson Mangum told me during a 2004 interview at her home in Ohio, "I think people mostly see [scrapbooks] when you're working on them. And [people] are there at your [place], you know, working on theirs."

At weekend crops in university housing, church camps, the back rooms of scrapbook stores, and the ballrooms of convention centers, practitioners have regularly stopped to display and talk about their creations, exchanging information about their lives in the process. Many crops build in opportunities for participants to share their work. Some groups, for instance, pause regularly to gauge project completion; this can be accomplished as a kind of game, where an object is passed "hot potato" style, and the person who's left holding the object at the appointed time must display her most recently completed layout or work-in-progress. Or scrappers may participate in formal display periods. Minnesotan Kristin Mathisen, twenty-six and the mother of one when I interviewed her in 1999, noted that at her club's meetings, "Everyone just puts out their finished books, and then people go through them. And if they like your page, they'll take a picture" of it. Child safety consultant and scrapbook sales representative Sandy Arnold remarked that her Creative Memories training meetings used to set aside time for unit members to share their albums; at the crops she held in a dedicated space in her own Indiana home, she encouraged "people just to walk around, and look at other people's albums." Display is often of this informal variety, characterized by browsing and interaction as people work.

Display also takes place *after* returning from scrapbook events. Kristin will sometimes wake her husband to show him what she's done after a long crop night. He's not always genuinely enthusiastic about her work while half asleep, but at more convenient times, she reported, "he always says, 'Thank you! You're doing such a good job, and this is really important'—so." Similarly, Troba's teenage son noted that one essential part of his mother's semiannual scrapbook retreat was gathering the family (her husband, four sons, and a daughter) for a presentation ceremony. Every time she returned, he told me, "We have to look through the um, pages—that's always fun to see." He was able to recall specific ways that her style and technique had changed over the years, contrasting her first attempt (what Troba recalls as a "gimmicky" layout involving a photo cut into the shape of a fish) with more technically and materially sophisticated pages.

Commercial events offer even more dramatic opportunities for "going public"; like the Show-and-Tell meetings of quilt guilds, these group events allow extrafamilial opportunities for expressive display (Langellier 1992). Performances of this sort can take place among tablemates at crops or workshops. As I walked through a North Carolina convention crop in 2004 and talked with some of the hundreds of participants, one young woman told me she thought scrapbook events were a great place for the "outgoing types"—the cheerleaders, class officers, and so forth—to meet each other, because those who attend these events are often thrust into close contact with strangers.

Or display can happen in front of a classroom. At the same 2004 event, when I asked Shannon Jones why she liked lecturing on the convention circuit, she said, "I don't know—I feel like it's pretty low key, unless you're actually teaching. Then you're pretty pumped up. [To Donna Downey, a fellow teacher:] I know *you* are. You're *jazzy.*" Donna laughed and agreed, realizing that because she's an only child, "all my *life* it's only been the Donna Downey Hour." Surprised to think of her lecturing as a performance, Shannon realized, "See, and I like musical theater, and that's my favorite hobby. I *love* to perform." Both women, in fact, engage in high-energy, expressive interactions with the women in their mobile classrooms.

Performance, and the evaluation that always attends it, may also be more formal and anonymous. For example, most state and county fairs have added scrapbooking as an exhibit class.[18] The extension agent at one midwestern fair told me that layouts are judged in terms of their neatness, balance, workmanship, photos, and attention to narrative content. (Correct grammar, she whispered, often tipped the scale in favor of a champion.) Yet even layouts that don't earn a ribbon are still displayed. Consumer-oriented conventions also generally sponsor a series of open-entry contests, and all submissions are posted on walls throughout the weekend (Fig. 2.10). These decontextualized pages function as exhibits explained in part by whatever journaling the maker has opted to include (the framing of these contests often encourages personal and emotional outlays), and passersby study them intently. Standards for evaluation sometimes privilege narrative content over technical skill: in contests of these sorts, even mediocre design or blurry photos can be trumped by a sincere and thoughtful story.

The internet is another site for public display and evaluation. In chatrooms, scrapbook practitioners regularly discuss and critique the work of industry professionals, who check in with these sites to see what people are saying about them. Much like the electronic venues through which indie crafters display and sell novel embroidery

FIGURE 2.10.
Some of the nearly six hundred participants at a three-day Creating Keepsakes University weekend (this one at a hotel in Provo, Utah) browse the contest entries displayed in May 2006. (Photo by author.)

patterns or other hip handmade forms,[19] dozens of online scrapbook galleries existed at the turn of the millennium; anyone could upload scans of a layout and wait for comments. The largest magazines and online sites regularly monitored these posts, highlighting exceptional offerings as part of official "layout of the day" features and offering standout creators employment as product designers (or a chance to move their work into a print publication).

Which brings us to a pinnacle of display venues: publication in competitive magazines and "idea books." Before the economic bubble burst and digital publishing rose to the fore in the mid-2000s, print publication represented the least ephemeral, most closely regulated, and quite lucrative option for scrapbookers. Each of several popular magazines—including *Creating Keepsakes, Simple Scrapbooks, PaperKuts,* and *Memory Makers*—was filled with layouts submitted by "regular people," as well as spreads and columns designed by those who had become industry personalities. In 2004, Shannon Jones, an Arizona native, mother of three girls, and student of elementary education, political science, and musical theater, explained how she got her start in the industry:

> And I was just sitting around scrapbooking with some friends of mine, and we were looking at [a] magazine, and I said, [pointing to it]
>   "I want to be in here."
> And we all laughed, and—especially *them* [laughs]—and I said,
>   "Yeah, I'm gonna do it with this page, right here."
> And I worked on it and worked on it, and sent it in for a little contest to win a scanner.
> And, I got a call from [an editor], and she said, "This is fabulous, we're gonna bump the person who was supposed to be in this slot and we're going to put yours in and it's coming out next month!" (Christensen 2004a)

Shannon—whose mother was a professional photographer—soon became a regular magazine contributor, was asked to publish a book, and started teaching at national conventions. Her story is not unique.

The benefits of going public, then, include broadened social networks, professionalization of kinwork (Di Leonardo 1987) and other forms of lay skill and knowledge, and redefinition of what counts as publicly valued labor. Not incidentally, reframing care-work through reflexive documentation has also led many scrap-bookers to an increased awareness of their role in adding value to the lives of the Others they document (cf. Clifford and Marcus 1986). But the endeavor carries risks as well: visibility invites greater scrutiny (and the power differentials that are built into fame and critique), reduces intimacy, and potentiates a greater range of possible interpretations.

Reframing Hierarchies: Maintaining Rapport in Public

Even as they seek greater recognition, then, scrapbook makers also adopt verbal and visual strategies to manage access and maintain preferred meanings. Talk and commentary is carefully tuned to minimize the risks that attend public performance, whether they involve bruises to the personal ego or the alienation of an individual from the larger group. Though these strategies may be motivated in part by a desire to avoid censure, the dialogue that emerges during group meetings and display events also fosters mutual recognition of similarities and differences among participants, an awareness that can trouble generalized assumptions about what it means to be a woman or a mother.

As we have seen, the transformation of intimate kinwork into embellished artifacts can lead to both visible exposure and monetary gain. Some scrapbookers crave the validation engendered by design competitions and magazine publication; one industry insider, for instance, remarked that many contesters "*want* recognition. And they *want* to know that theirs is *innovative* and the *best.*" But display also has drawbacks; for example, a focus on winning can be perceived as presumptuous and off-putting. She remembers (and not with fondness) one woman who introduced herself at a board meeting by reportedly saying "Well. How many times have *you* been published?" Is it possible, then, to go public without legitimizing the rankings that result from evaluation? And can a woman put her soul—or talent, or daily life—on display without being branded selfish, immodest, or indiscreet?

Yes, if these activities and accomplishments are discursively framed as "more private." To this end, scrapbookers-in-public often carefully downplay their achievements or present them as collaborative accomplishments. That is, evaluative commentary among scrapbook enthusiasts is overwhelmingly supportive; in addition, although individuals may distinguish their own style from another's, they do not, as a rule, rank such differences in a vertical hierarchy.

Many scrapbook celebrities, for instance, take care not to ground their personal authority in publications or institutional credentials. One instructor expressed annoyance that the few men in her line of work seemed always to be talking about their art or graphic design degrees; to her, these appeals to authorizing documents constituted transparent bids for power. During my fieldwork period, industry professionals tended instead to position themselves as every-women, regularly minimizing claims to unique ability or authority. In the late aughts, the "About Us" page of an online instructional enterprise called Big Picture Scrapbooking confided "We don't really like corporate-sounding resumes"; instead, each teacher was represented by a casually posed head shot and a personalized top-ten list. Prompts included "Ten 'essentials' I cannot be without," "Nine words I love," "Eight bands who keep me company on any given day," "Seven things I love about my 'every day' life," "Six places I'd love to visit before I die," "Five things I do every day without fail," "Four stores/websites I frequent," "Three photos I love," "Two decisions I've never regretted," and "One additional and very important thing you should know about me." By 2015, Big Picture Scrapbooking had become Big Picture Classes, but the site still offered intimate portraits of teachers, linking visitors to personal blogs and including video interviews that emphasized accessibility, informal learning, home work spaces, individuality, haphazard children's bedrooms, and off-the-cuff conversation.[20] In these spaces, institutionally trained scrapbookers are placed on an even footing with those who have acquired their skills in other ways.

Another strategy for reframing public visibility is to stud commentary with self-deprecatory disclaimers. Since performance in its public and masterful senses has for so long been marked as inappropriate for women (e.g., McLeod and Herndon 1975; Sawin 2002), those who do choose to act artistically in public space may deny that their actions count as performance at all.[21] For example, I often found that when complimented, scrapbookers would quickly mention projects gone awry or enumerate skills they did *not* possess. One scrapbook consultant I spoke with explained that she was "addicted to colors of fabric" and saw scrapbooking as a way to create art "with glue and scissors—instead of [with] a sewing machine and instructions that I don't understand!" Similarly, when I asked Troba (a precision quilter with a keen eye for striking color placement) to talk about what she thought made a good page, she played up one of her early "failures" instead (Figs. 2.11, 2.12):

FIGURE 2.11.
Troba Nelson Mangum prepares dinner while we chat at her home in Columbus, Ohio, in December 2004. (Photo by author.)

FIGURE 2.12.
"And THEN THERE'S CRAIG!! Springing *forth from the stream!" The
infamous Fish Layout, the first scrapbook page Troba Mangum created,
ca. 1994.

> I think when I look in my scrapbooks I get a great, laugh, out of them.
> Because I kept the very first, um, one I ever did?
> The very first page I ever did [in the early 1990s]? Which is . . . so awful.
> I even cut a picture of Craig, in a stream, into the shape of a fish.
> [laughs]

Later in our discussion, when we were talking about compositional
processes, she returned to this spread. Her initial comments about
composition claimed special skill in the form of intuitive knowledge,
but then she immediately downplayed that expertise. When I asked
how she could tell when a page was complete, she responded,

Just how it feels when I look at it.

You can feel if something feels finished? You can feel the holes, um—
and sometimes you're totally wrong! I mean you think this is a great
page, and you come back and look at it a month later and you have to
laugh your guts out, because it's so bad!

Ergo, the fish page. [laughing]

It was so bad! I just love it! It's so nasty! (Christensen 2004b)

In another and more overt disclaimer, Danielle Denise France Todd,
a bank sales manager from Evansville, Indiana, characterized herself
in 2003 as "not real creative"—just before admitting that adapting
ideas from others allows her to look back on some pages and think,
"'Hm. That's kind of cool.' You know?"

Others minimize their expertise by demystifying it (see Baudrillard
1968). Some point to models they've built upon, crediting layout
designs, particular techniques, or journaling prompts. Traditions of
"scraplifting"—reproducing or modifying the layouts of others—and
tip-sharing are in fact explicitly encouraged by magazine features and
crop-event structures, and they reinforce a sense of collective performance rather than individual genius. In addition, a number of practitioners I spoke with were quick to explain techniques that used
specific tools, placing their creations within a "workmanship of cer-
tainty" rather than one that emphasized risk (Pye 1968). Thus, when
I complimented Kristin on some page accents she'd made, she replied,
"Oh that was easy. It was just a circle punch."

Women who have become commercially successful are also apt to
brush aside intimations of exceptionality. Donna and Shannon characterized themselves and their work as "a dime a dozen"; for every one
person who makes it as a celebrity, Donna said, there are another hundred or so people who are "*all* equally if not more talented." Both
women attributed their achievements to timing. When I asked another
maker (whose "public" layouts were regularly featured in magazines)
what made a good page, she responded with an admission that good
design—design that follows compositional rules—came naturally to
her. But she characterized this gift as a haphazard and ultimately unnecessary one—unless, of course, one wanted to be published (an option
she dismissed as an important goal). In a related move, prominent
lecturer Stacy Julian worked to establish rapport with an audience
in one of her workshops by praising a colleague, then lumping herself
in with "the rest of us" who aren't intuitive in the same way—even

though her first book, a self-published twenty-seven-page volume called *Core Composition* (Julian and Darcey 1997), is an effort to define the principles of design that she'd been using since her first forays into the practice.

As I asked scrapbook makers how others encountered and reacted to their work, I noticed that another recurring strategy for minimizing overt claims to performance was reporting truncated versions of praise, rather than asserting personal talent directly (cf. Sawin 1992, 2002). Over and over, my consultants *quoted* other people's evaluations of their work, using what linguists call "direct discourse." Explaining in 2005 that she gets "so encouraged in scrapbooking," Bloomington, Indiana's Cherie Shields-Wilson recounted specific compliments she has received for her work:

> I hear people saying
>   "Oh!!! I've heard about your scrapbooks! I wanna see em! I wanna"—
> You know. (Christensen 2005a)

In a laughing aside, she added that her children's friends drool over the books she's made. A former art teacher, her reputation for being a talented scrapbooker had spread through word of mouth; after all, she noted, the books themselves are "inside your house, it's not like you *parade* them, or anything, you know."[22] Troba also employed reported praise when describing her motivations; additionally, she expressed her own surprise ("amazing") at the extent of the positive response she received:

> I like the color, I like doing it my way—I like it when people praise it, you know,
>   "You do beautiful work"—
> but you know what the most amazing thing is?
> To see your college-age sons *drag* friends to the couch, with their book
> *Danille:* [laughs]
> *Troba:* and show them, how cute they looked, when they were in kindergarten.
> And then, everyone is so . . . enthralled with them.
> Everyone! loves the books.
> The kids love the books, their friends love the books, [my husband] loves the books.[23] (Christensen 2004b)

Similarly, when recounting how a commemorative album was received, historian Betty Bridgwaters noted that her friends couldn't believe their eyes (Fig. 2.13). The album celebrated the life of June

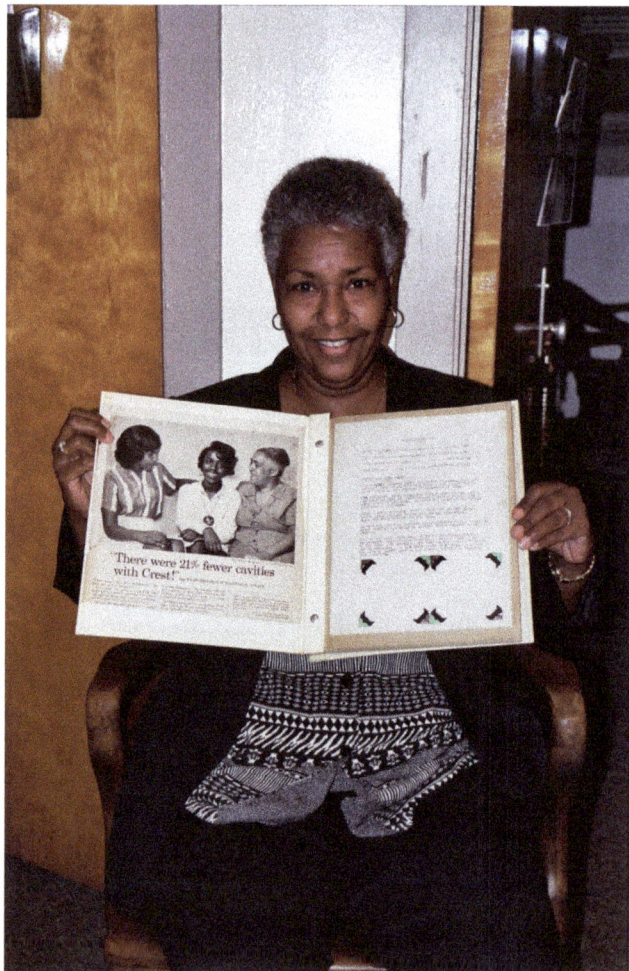

FIGURE 2.13.
Elizabeth Ann (Betty) Bridgwaters (Bloomington, IN) displays one of her adolescent scrapbooks in September 2006. This page documents her family's participation in the original Crest fluoride study. (Photo by author.)

Hammond, who mentored the Cosmopolitan Club, a Bloomington, Indiana, youth organization formed in 1959. Group members had kept postbound scrapbooks in the 1960s to document club leadership and activities; the books were displayed each year at the state convention of the National Association of Colored Girls Clubs. But the gift album for Ms. Hammond utilized all the materials and decorative resources available in 2003 to good advantage. When she showed fellow "club girls" the pages in progress, Betty remembered, "They would say,

'Oh, June's gonna love this.'
'Oh this is wonderful!'
'Oh this is so pretty.'
'Oh, I want one of these!'" [laughing] (Christensen 2003)

Likewise, Joan commented that she thinks people do appreciate her albums, recalling that she frequently gets "comments like, 'Boy *that* is really *clever*' or 'You did a good job on this page.'"

Stacy provides a final example of the ways discursive humilities can mitigate the potential for being seen as too publicly performative, too proud, or self-important. When I asked her in 2006 how she thought people perceived what she did, she said it depended on what circles one ran in:

If it's someone who's not a scrapbooker, they're like
"Really? They *do* that? There's people who have magazines just about scrapbooking?"
Versus people in the industry who view me as a celebrity—very weird.
To me, I don't get it—it's never been . . .
Oh, the person who bought our home. We're moving! The person who bought our home—
*We went over to talk to them about ripping up the really ugly stained carpet—cause I'm so embarrassed they might see how bad the carpet is when they move in—
and she said,
"Why didn't you tell me you were famous?"
And I go,
"I'm not famous." [laughing]
And she goes,
"Yes you are!"
She goes,
"My *co-worker*—I was tellin' her that I bought your house.
And she goes,
'*You* bought *Stacy Julian's house*?!! *Oh my gosh!* Do you know what this means??!!!'" [laughing]
I mean, and she goes
"I had no idea! What do you do?"
And I explained it to her. So
Danille: Is she going to *frame a square of the carpet? [laughing]

> *Stacy:* *Yeah! Exactly, yeah, tell her "you can have a big chunk of the stained carpet!"
> But you know—so, it just doesn't—I don't get it.
> You know?
> It's OK. I'm just me. I'm so lucky that I get to do what I love. And it makes me happy. (Christensen 2006)

By sandwiching reports of praise between accounts of being unrecognized by most people (and having stained carpet), Stacy acknowledged her influence without asserting it directly; furthermore, she characterized scrapbooking as a circumscribed sphere, a more private way of going public.

If one strategy for reframing public display is to minimize self, another is to maximize others. An excerpt from a 2004 *Chicago Tribune* article is telling in the way that two women relate to each other at a crop held in a local scrapbook store:

> Wearing a long-sleeve Ohio State T-shirt, [Maggie] Workman is laying out photos of a daughter's music recital.
> "The documentation is important to me," says the thirty-four-year-old Workman, who lives in Waukegan [Illinois]. "I'm not as much of an artist as these guys."
> "You hold your own, Maggie!" [Noreen] Biegalski pipes up from across the U[-shaped workspace].[24]
> "I'm a scientist," Workman says with a shrug.
> A PhD in environmental science, Workman is an instructor at DePaul University in Chicago. Her husband is a biochemist at Abbott Labs, and they have eight-year-old and twelve-year-old daughters.
> "I try to come here once a month. It's the only time I scrapbook. I don't do it at home," Workman says. "There's the quilting bee aspect [to the crop]. We talk a lot."
> Motioning to the scrapbookers around the U, Workman says, "You know everything about them—from the pictures." (Reardon 2004)

Maggie Workman's assertion that, as a scientist, she prefers documentation over decoration could be read as an alignment with regimes of value that privilege the classical and the evidential. But her comments, and the enthusiastic encouragement offered by Noreen Biegalski, also point to other discursive and evaluative norms of scrapbooking in public. In addition to minimizing one's own skill or abilities, scrapbook practitioners often downplay the power differentials of performance by emphasizing collaboration and strenuously avoiding overt criticism of others (cf. Goodwin 1980a, 1980b; Kalčik 1975; Maltz and Borker 1982).

Like Workman, Kristin prefers to work on scrapbooks with others because, she told me, being in company turns scrapbooking into an occasion and motivates her to "get stuff done." Amy, when describing her preferred context of creation in North Carolina, emphasized material and ideational collaboration. She told me she preferred to scrapbook with Michelle—"I mean, I *will with other people, I'm not gonna be *mean," she told me—but if they're not sitting together, Amy will seek her out to ask, "Michelle: What colors do I *need to go with . . ." Michelle described the two as "necessary partners" who often get together at home and "just take over a room." Likewise, Kristin characterized successful pages as mutual efforts:

> I'll hold up a page and I'll say,
> "What does this need?"
> and [her friend Mandy]'ll say,
> "Oh do this,"
> and I'll do that, and I'll be like
> "Oh, it's *great!" (Christensen 1999)

Here, and like so many other women with whom I've worked, Kristin uses quoted dialogue to verbally model (and therefore underscore) the back-and-forth exchanges that take place during crops, and she assigns credit to Mandy for discovering the all-important final flourish. Troba, too, situated her own skill in relation to the assistance of others. When I asked her to explain how she identified visual gaps that needed to be filled, she protested,

> I can't. I can't explain it!
> I just, I just look at it, and if it's right, it's right. If it's wrong, it's wrong.
> And sometimes it feels right, but then I'll say to somebody,
> "You know, what do you think about this?"
> —and they'll add something, and it makes it better.
> It's not like it's [an] *infallible* [gift]. (Christensen 2004b)

Collaboration thus takes place during creation, as well as during narration and interpretation. Women I've interviewed tend to punctuate their speech with interrogatives (rising inflections at the end of sentences) or with what might in other contexts be called influencies or fillers, such as "you know." In fact, these interjections are phatic in function (Jakobson 1960): they seek to establish and maintain contact with interlocutors. Stacy's repeated "OK?" during a lecture I recorded at a Creating Keepsakes University event is a recurrent call that seeks a response, a discursive "search for consensus or affirmation" that a "viewpoint is not subjective, but arises from shared experience" (Yankah 1995, 14). Inviting the affirmation of the audience is a way of proclaiming solidarity and projecting humility while simultaneously asserting one's own perspective as correct, or at least broadly representative.

Another way that scrapbookers mitigate the hierarchical underpinnings of display is by avoiding direct criticism. Ellen Scott, an administrative assistant and single mom from Columbus, Ohio, noted in 2004 that scrapbookers had an "unspoken bond"—one forged through shared experiences and activities, but also reinforced through discursive practice:

> And you don't *criticize?* You don't criticize!
> I mean, there's, like Leslie has said to me—especially on my first album I did, when I didn't know what I was doing—and now that I kinda think I know what I'm doing I wanted to go back and redo it. And she has said,
>> "Absolutely not! That's your first album you did! You want to keep that to see where you've grown!"[25] (Christensen 2004b)

Positive peer reinforcement like this is the norm in online galleries, where TFS, "thanks for sharing," often shows up in posted comments. In 2004, I interviewed Renae Call, a fifty-three-year-old small-business owner and mother of three who had lived in Utah for much of her adult life; she noted that online or in person, "everybody's always *complimenting* everybody on their *page*"—even if "inside you're saying, 'Eww!'" she laughed, "*You're *still* usually *nice."[26] Fortunately, the multiple semiotic options at play on a scrapbook page provide a way to be tactful without being completely disingenuous: if aesthetic or technical excellence appears lacking, viewers can always praise photographs or comment on a personal connection to a page's content. One unusual message board post responded to a layout about someone's DH ("Dear Husband") by offering a specific criticism—but even then, this observation was couched in visibly supportive language (and punctuated by exclamation marks). The writer, who had posted 2366 times since she joined the board in August 2004, commented in March 2007:

Love the colors! He's my favorite "little boy" kind of cute and I'll bet he
has a happy expression even when he's not smiling. Lucky you!
You didn't ask for it . . . but I would want to know . . . the word "patient" in
the middle of your journaling has the "i" and "t" reversed. I do things like
that myself often enough! My most popular mistake is leaving the "y" off
of "they"!

Her use of ellipses warns the reader to prepare herself for critical
feedback; at the same time, "I would want to know" establishes her
apparently nonhostile intentions and her volunteered personal foi-
ble attempts to keep the interchange on equal footing.

Finally, to avoid claiming excessive authority, scrapbook makers
often couch their criticisms in terms of personal preference, rather
than absolute truth. Susan considers scrapbooking nurturing both
for herself and for her family and contrasted her hobby with other
leisure-time options. She thinks to herself:

> "Wow, what a bonus, what I enjoy is something that also blesses my
> family."
> Instead of,
> "Wow! You know, what *I* enjoy is, surfin the internet" [laughing]
> you know—
> some,
> and not that that's a bad *thing*, but,
> I just consider myself *doubly* blessed that, something that brings *me* joy
> can, down the *road,* bring my family, joy.
> So. (Christensen 2003)

Time and again, my interlocutors indicated that they carefully mon-
itor their speech, catching themselves when they teeter on the verge
of disparagement. Like Susan, as Sandy (a distributor for a direct-
sales scrapbook firm) explained how she favored handwriting in her
albums, she contextualized her preferences in relation to other
known options: "And not that printing it out on the computer is
bad, you know, but just that—I always try to encourage people to, at
least write *some* [by hand]." In another example, Troba told me that
she thinks basic identifying captions on a page are important, "but I
don't write *long*, poetic *prayers* to my children or anything like that."
Then she broke off her comment in order to qualify the criticism
implied by her mild sarcasm, characterizing her own efforts as probably *too* simplified:

these people [who do write "poetic prayers"] are just gifted. [laughs]
*They just write the most amazing things on those things.
And we just say things like,
  "We are so proud of you today, *Craig."
[laughing]
  "Look at you! You got this big award!" (Christensen 2004b)

Similarly, after telling me that she didn't do digital scrapbooking, Troba followed up by saying "I think digital is probably just totally awesome—but that doesn't represent me, in my work. At this point."

Regularly positioning one's work and one's opinions in this way requires acknowledging others and admitting the validity of their choices. After Shannon contrasted Donna's "neat, tidy, and graphic" style with her own "super feminine" approach, she immediately proffered this explanation: "But I have three girls, too, and that probably has a lot to do with it." The differences between the two women's styles, Shannon implied, had to do with personal experience (and broader social norms) rather than functioning solely as an indicator of taste—or the lack thereof. In these spaces, then, women speak publicly, claiming attention but not claiming it exclusively.

Beyond Face Value: Disclosing Secrets

Going public has risks that extend beyond perceived arrogance or potential censure; control over access to specific content is a concern as well. Album creators manage addressivity—the "quality of turning to someone" (Bakhtin 1986, 99)—in different ways as they orient texts toward particular publics.[27] As noted above, some makers foreground the risks of potential misreadings and fashion their books using explicit, "finalized" written narratives (Bakhtin 1986) that spell out preferred interpretations and ensure that authors get the last word in the future. But there are times when the possibility of extended publics also encourages strategic gaps and silences—moves made possible by specific formal choices and contexts of display.

For example, scrapbook leaves can be re-ordered, the contents of their pages can be changed or rearranged (an enduring advantage of adhesive "photo corners"), or entire pages may even be strategically withheld. When Anne Cohen was interviewed by Smithsonian fieldworkers in the 1970s, for instance, she referred to photo pages featuring a secret boyfriend as "my removable past": Cohen hid this evidence away when she showed her parents her albums (Zeitlin, Kotkin, and Baker 1982, 193). Kendra Barr also held some pages in abeyance. An administrative assistant at a large midwestern corporation, she had fled an abusive relationship, changed her name, and raised her children alone.[28] In 1999 she told me that she had visually documented some events related to her former husband but was waiting to display them until her two preadolescent boys could "handle it," when they were mature enough to "stand on their own two feet and not have to pay for their dad's mistakes."

Other scrapbook makers cultivate the mystery of what Barbara Kirshenblatt-Gimblett has called "the split sign": a situation in which viewers can access the signifier, but not necessarily the signified (Kirshenblatt-Gimblett 1998); there's stuff to look at but very little obvious "story." Joan's Thanksgiving layout, for instance, simultaneously invites viewers to fellowship and effectively makes the pain of that hospital experience available only to those who care enough to ask.

In general, split signs are sites of social influence: riddles, proverbs, magic tricks, tall tales, pranks, and the like all depend on the reconciliation of signified and signifier, sometimes manifested in a theatrical revelation that underscores the power of those in the know. These "little drama[s]" of secrecy direct audience attention and perception by "regulat[ing] the rhythm, pace, [and] range" of what is shown, and what is not, when, and to whom. Sometimes value resides in the split itself—as in what Kirshenblatt-Gimblett calls the overcoded signifier, the form that "makes a show of hiding" (Kirshenblatt-Gimblett 1998, 255). That is, the appeal of things designated *exotic*, the power of deliberately esoteric languages (pig latin, carney) or secret family recipes, and the authenticity of carefully guarded slack key guitar tunings all lie in their trumpeted impenetrability to the casual outsider.

The material characteristics of scrapbooks—tangible layers, multiple surfaces, diverse materials—tend toward overcoding, allowing makers to run riot with meanings. This is intertextuality you can touch, a kind of chaos rich with sensory and narrative possibilities that are obviously difficult to make sense of.[29] It is the very opaqueness of "found" (maximally decontextualized) scrapbooks, for instance, that makes them so attractive as *objets d'art* (see Helfand 2008). Alternatively, the existence of secrets can be broadcast by peekaboo objects, such as the large paper tags that became popular among scrapbookers several years ago because they could be inscribed and tucked behind photos and other page items as a form of hidden journaling. These objects materially structure levels of personal disclosure and audience engagement, and they are not necessarily subtle about it.

But another kind of "asymmetrical split"—the *under*coded signifier, the "flat" surface that seems to hide nothing—carries a more nuanced kind of power. In this mode, pages can easily be passed over as "trite" or "formulaic"; gatherings of women can be waved off as "parties" or "getaway weekends." Undercoded signifiers are self-protective shibboleths, dividing insiders from outsiders in a way that avoids proclamations of secrecy in favor of an easy readability that deflects further scrutiny (Kirshenblatt-Gimblett 1998, 255). Because such apparently transparent forms can be dismissed as "unimportant, innocuous, or irrelevant," the undercoded signifier has long been a useful tool of the subaltern, mobilized through tactics that include juxtaposition, indirection, trivialization, and professed incompetence (de Certeau 1984 [1988]; Radner and Lanser 1987, 420). Thus, although masking or concealing information *is* part of contemporary scrapbook practice, it need not be evidence of unconscious repression or willful naïveté. Instead, building in the option of oral elaboration allows a book's maker to reveal information at her discretion, controlling who has access to pain, trauma, disappointment, or other types of intimate emotion and knowledge.

I would argue that this kind of strategic management is not new to scrapbooks and photo albums. Despite the semiotic complexity of these forms, they are often taken at face value, so that "typical" album pages—vacations, holidays—may be interpreted in terms of photographic convention or sanitized reminiscing. Richard Chalfen, for instance, has characterized snapshot collections as proud expressions of "conspicuous success, personal progress, and general happiness" (Chalfen 1987, 99). Katriel and Farrell write, "The kinds of stories [scrapbooks] typically tell, their interlacing of particular traces of 'good times' in a macrotext conveying a sense of the coherence, the wholeness, and the significance of the individual life, all suggest a distinctly American notion of what 'the good life' and the telling of it ought to be" (Katriel and Farrell 1991, 15).

Certainly, the bulk of amateur photography involves deliberate selection and editing and appears decidedly optimistic, a factor that has led several scholars to point to family albums as false or oppressive constructions (e.g., Spence and Holland 1991).[30] Others, as in *The Onion* story quoted above, see the act of "layering carefully chosen photos with brightly colored paper, elaborately patterned borders, and whimsical stickers" as an exercise in psychological denial. When I spoke with Jane Bauer, a university professor in her late thirties, on a fall day in 2003, she concurred with these perceptions, remarking that genre and social norms conspired to make her produce layouts that concealed more than she wished.[31] Jane kept a set of album-compatible "calendar pages" on which she jotted down milestones as well as everyday events, such as when she was sick. But she agonized about reconciling her private reflections with the expectations of an unknown and public audience when it came time to label her scrapbook pages. She wished she could be "more self-disclosing," as she would if she were writing in a journal. "[B]ut I can't, because this is for, you know, public viewing."

> *Jane:* I'm afraid to do that, because I don't wanna . . . be embarrassed, and so what I end up writing sometimes feels kind of hollow.
> It's like what they show in those *Crop Talks* [sample layouts distributed at Creative Memories workshops], you know—the sorts of vacant things that are written as examples? You know,
>> "Oh Grandpa Charlie always knew how to,"
> you know,
>> "make us laugh!"
> and stuff like that *[laughing]* just so—
> But that's what you end up writing, because you don't wanna, be negative, and you don't want to be offensive, and you don't want to make your kids uncomfortable, because you really were *[laughing]* angry at them—
> *Danille:* "I can't believe you did this!"
> *Jane:* "You're really *angry*, Uncle Charlie!" *[laughing]* or whatever. [. . .]
> Sometimes I imagine, like, my kids grown and they're married, and their spouses are looking at [a scrapbook] and thinking,
>> "God! What a Pollyanna! Was she, ah, you know, so threatened by the truth that you know, she was afraid to admit . . . ?"
> And I am like such a, you know, so *willing* to admit *[laughing]* when things are not good, it's not my personality [to be a Pollyanna], so sometimes I read this and I think,
>> "That's not—that doesn't *sound* like me!" *[laughing]* (Christensen 2003)

Jane's concerns about flattening her experience and misrepresenting her personality coincided with public commentary—books, newspapers, magazine articles—that decried or dismissed as simplistic the positive images and mass-produced objects in scrapbooks at the turn of this century (Christensen 2011). But these critiques fail to recognize that a given scrapbook is a stretch of discourse made unique both in terms of its formal characteristics and "the infinitely rich specificity of the context in which it is embedded" (Urban 1996, 21)— contexts that include lived experiences, personal associations, appended oral commentaries, and the photos themselves.

Photos are complicated, for instance, by the fact that they act as both icon (literally representing a real-world referent) and index (evoking a range of related forms and feelings).[32] Michelle's pages, for example, often emphasize the latter. While explaining how she justified her "non-chronological albums," Michelle told her friend Amy and me how she composed pages by juxtaposing disparate texts and images, so that the accompanying written and felt narration might have little obvious connection to photos on the page. She'd recently attended a family reunion, the first in a long while:

> The part that struck me was when, we were all gathered around, all my crazy family members, and my um, daddy's oldest brother, said grace.
> And, I hadn't seen Lewis in *years*—he lives a ways away—and I just remember that, when he started saying grace, it was like cold chills broke out all over my body, cause his voice sounded just like *Grandpa*. And Grandpa had been dead for, my goodness, since I was, fifteen or sixteen.
> And I had *forgotten*, what his voice sounded like.
> But when Lewis said—it was like a *flashback*. I was like "Oh my G—" it, it just, it brought a memory back.
> And it just—cold chills. (Christensen 2004a)

Of everything that took place at the reunion, it was this vocal reembodiment of her grandfather that made a lasting impression on Michelle. But, she said, "I can't, don't really have a picture for *that*." Consequently, the images on the page—excepting one of her father and his brother Lewis—have nothing to do with her written journaling, which approximates the story she told me above.

These affective subtexts are precisely what make scrapbooks so valuable to the people who create and view them. And because the possibilities for expressing meaning are so varied in scrapbook texts and performance contexts, access to these underlying narratives can be controlled in several ways. While Michelle chose to write down the associations sparked by her uncle's image, others (like Joan) reserve narrative elaboration for oral contexts. Smithsonian Family Folklife Program researcher Amy Kotkin found in the mid-1970s that although photographs presented in albums were "almost invariably happy, many of the stories they rekindled were not" (Challinor et al. 1979, 260).[33] Between 1974 and 1976 Kotkin and her colleagues interviewed more than five thousand residents of and visitors to Washington, DC. Their findings led them to recognize the importance of viewing photographs in interactive contexts, encounters in which images are situated in a fuller interpretive surround.[34]

For instance, a snapshot of a mother and her grinning offspring gathered around a seashore restaurant table prompted one of the featured children to recall how her father had abandoned them just prior to that vacation; photos from the 1920s reminded another person that the family's treasured camera had to be sold a few years later, during the Great Depression (see Zeitlin, Kotkin, and Baker 1982). One woman explained "a jovial photo of her Aunt Rose on an American beach," by relating the aunt's traumatic escape from Russia, while a "proud photo of Nancy's father in his World War II army uniform was counterpointed with a tragic tale of how he had shot a little boy in Germany, mistaking him for an enemy soldier" (Challinor et al. 1979, 260–61). Similarly, when Martha Ross—who, in the 1940s, compiled the cheery documentary album called *Little Visits with the Rosses*—shared her scrapbook with Smithsonian researchers, she expanded on how small her first home was and revealed that she was in fact pregnant and feeling unwell when her husband snapped a smiling shot of her doing dishes at the sink (Zeitlin, Kotkin, and Baker 1982). Among my own research pool, Cherie explained a layout of her smiling adolescent daughters making Christmas cookies by commenting on how "*snippy*, and *grouchy*" the evening had become before it was over. Rather than longing to relive the event, she found the memory both humorous and important to depicting the realities of family life.

Gaps in written narrative, then, may signify or accomplish several things. As in ballads or folktales, "missing" information may assume an intimate audience for whom expository information would be unnecessary, or it may signal that overall aesthetic experience is valued over specific ("coherent") elements of content. Gaps highlight the ways that lived experience can never be completely represented by one or even many semiotic systems (e.g., visual photos, written language, meanings associated with color or texture, and oral speech). And as we have seen, elisions and nonnarrative means of communication effectively "code" cultural productions, offering protection and control to their creators by making them more or less impenetrable to unknown or unasked-for audiences (cf. Radner and Lanser 1987; Scott 1985).

Even verbally finalized scrapbook pages can hide information—in part because they are "inseparable from the human actors who know, remember, embody, do, and perform" them (Kirshenblatt-Gimblett 2006, 32). Layouts that do explicitly address heavy subjects in writing are often made more poignant in the course of an added layer of narrative, during oral performance. For instance, the biggest contest at the May 2006 Creating Keepsakes University (CKU) event in Provo, Utah, was sponsored by Chatterbox as part of that company's "Make It Meaningful" campaign. Entrants in an international competition were asked to craft a layout that addressed the prompt "I scrapbook because the best things in life aren't really about the things" and also write an essay to be submitted with the finished page. By the time prizes were awarded at the event's commencement ceremony, hundreds of women whooped in support of the contest winners, who then explained their personal backstories to the cheers, sniffles, and questions of the crowd. The narrators and their layouts evoked the challenges of dealing with teenagers, multiple sclerosis, potty training, widowhood, and single motherhood. Many of these stories were not new to audience members, who had already read captions and other abbreviated versions written on the layouts themselves. As a result, cheers and murmurs of recognition greeted the announcement of winners: "She's the one who came with her husband," "She included those photos taken by her disabled son." During the award ceremony, the judges extolled the content of these explanations—but they lamented the fact that little of that contextual information was represented on, and thus permanently linked to, the layouts themselves! The pages risked being understood only at face value.

The layers of public and private meanings made possible by scrapbooks in use are perhaps best illustrated by a final example, in which one scrapbook maker encapsulated a particular moment and its attendant knowledges for future audiences, using her layout—and its extensive written text—as a jumping-off point for oral elaboration and expansion in an intimate present. During a discussion in early 2005, Ann Russo mentioned that she believed contemporary scrapbooks and scrapbooking helped to combat "superficiality." When I asked if she could show me an example of what she meant, she turned to a page from her youngest daughter's album. Ann is a cheerful, confident woman who had spent significant time overseas as a military nurse and physician and was mothering full time when I interviewed her. The page she showed me—which she refers to as "the one where I try to convey how exhausting it is to be a mother"—details the exhaustion and reduced sense of self she felt while this daughter, her fourth child, was an infant.[35]

The layout—inspired by a mother/daughter silhouette she once saw in a magazine—is saturated with the deep reds and pinks that can signify both frustration and love; a series of mats draws attention to the single photograph at the page's center (Fig. 2.14). Mother and child fuse in silhouette, their indistinct outlines further blurred by the vellum overlay. Lack of handwriting and proper names on the page universalize the image: this could be any mother and child, despite the fact that the text accompanying it is extremely specific in address:

> There's nothing quite like the sheer exhaustion that follows having a baby. No one can prepare you for it, and no one can adequately describe it. You just have to experience it for yourself. Even though you were my fourth child, I still found the fatigue relentless. I spent many hours in the rocking chair, dozing as I nursed you. I recall one night when you had been up most of the night. I was at my wit's end—I'd tried nursing you and rocking you, but you just wouldn't settle down. Finally I awoke Daddy and asked him to help me. We both sat in your room, alternately trying to calm you down. Somehow we got through it, although I don't exactly remember how . . . [all ellipses in original]

> The first three months were such a blur. My body ached from fatigue, and I found it hard to concentrate. One evening, when you were two months old, I took Carolyn [her older sister, six or seven at the time] to a scrapbooking crop. I had such a hard time carrying on conversations with the other women—it was like my mind was in a fog, and my whole body was moving in slow motion. At check-out time, I found I couldn't even do the simplest of math calculations! Even after you started sleeping through the night, I still didn't feel quite "right." I just didn't seem as quick-thinking.

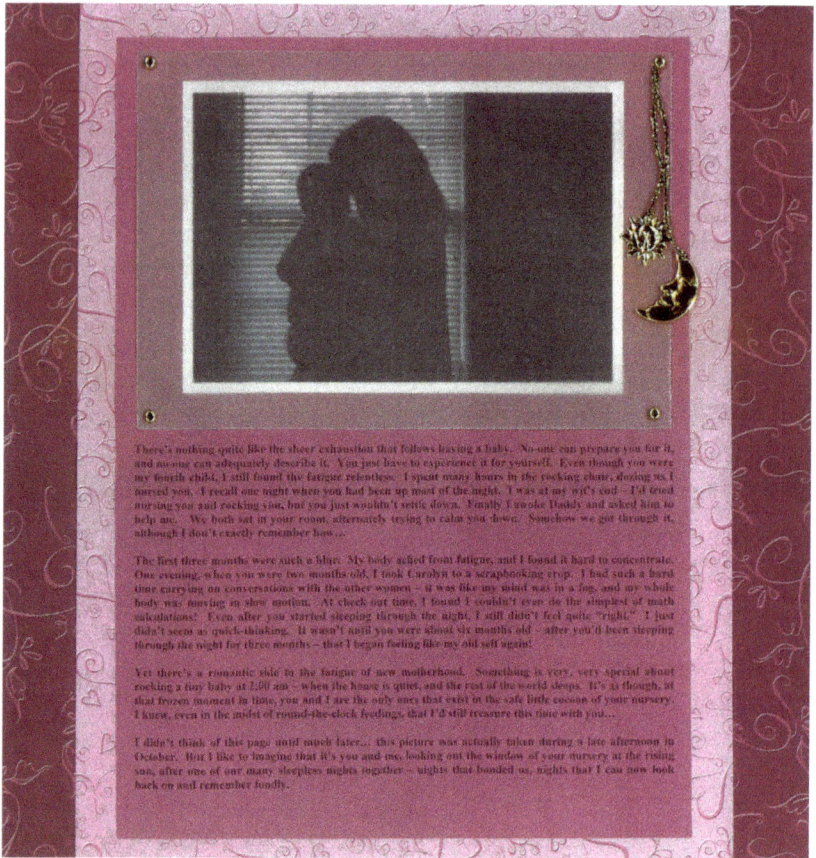

FIGURE 2.14.
Ann Russo (Columbus, OH), "New Motherhood," twelve- by twelve-inch patterned paper, ca. 2003.

It wasn't until you were about six months old—after you'd been sleeping through the night for three months—that I began feeling like my old self again!

Yet there's a romantic side to the fatigue of new motherhood. Something is very, very special about rocking a tiny baby at 2:00 a.m.—when the house is quiet, and the rest of the world sleeps. It's as though, at that frozen moment in time, you and I are the only ones that exist in the safe little cocoon of your nursery. I knew, even in the midst of round-the-clock feedings, that I'd still treasure this time with you . . .

I didn't think of this page until much later. . . . This picture was actually taken during a late afternoon in October. But I like to imagine that it's you and me, looking out the window of your nursery at the rising sun, after one of our many sleepless nights together—nights that bonded us, nights that I can now look back on and remember fondly.

The page, unlike some of Ann's other layouts, is unsigned, a choice that (coupled with her use of first-person pronouns) suggests that Ann does not anticipate that this page will "wander" (see Glassie 1989, 184). It is a self-contained unit of discourse, framed by a thesis and a conclusion and filled out with sequenced episodic reflections and semantic and thematic parallels. But the ellipses suggest there's more to the story. And, indeed, as we spoke together Ann volunteered additional information:

> *Ann:* BUT: I wanted to get the point across, that there are, that the *fatigue* of *motherhood* is just—you know, those early months, it's just overwhelming!
> And, I guess part of it is, to, when my g, my *kids* have kids, to tell them,
> "It's OK. You're gonna get *through* this"—
> you know,
> "It's *normal* to feel this way."
> Because when I had my first *child*, um, I had a *terrible* time.
> Um, I had an emergency cesarean, and I almost bled to death and, um, typical me, I thought I was gonna do it all on my *own*!
> I said,
> "OK, I've got six weeks off from work, I'm gonna get all my lectures done, and d-d-d"—
> and *boy*!
> I was *blown away* by how—it's a very humbling experience. You are *not* in control.
> And, in retrospect, I realize I had a pretty severe postpartum depression.
> And, I had *no* help at home—my husband had gone back to work, I didn't *want* anybody to come in, I wanted to "*bond* with the *baby*"!
> *Danille:* Uh-huh.
> *Ann:* And, I wish somebody had just said,
> "Ann. No. *You* need somebody to take care of *you*."
> I was *so* anemic, I couldn't even stand up to make myself a sandwich.
> *Danille:* Oh my—
> *Ann:* I couldn't stand up.
> I couldn't even make it down, to the road, to get the mail.
> I couldn't make it down my *driveway* without stopping and resting twice.
> *[lets out an exclamation of air]*
> And here I'm trying to take care of this infant.
> It was *horrible*.

And—if somebody had just *told* me, you know,
    "You're gonna get through this, *accept* some mothering from some-
    body else"—
you know, but we don't have that kind of *culture* anymore.
We, we're expected to be *off* on our own, and be, *supermoms*, and I think
it's a *very* um—it's very *damaging*. [pause]
And so part of it—I want to show them,
    "It's OK, and this is how I *felt*." (Christensen 2005b)

This page, and the interactive contexts it encodes and generates, offers Ann the opportunity to do just that—to tell her children by means of print, image, and conversation the knowledge she wished others had shared with her. But her most poignant social critiques and moments of self-reflection are reserved for the oral telling, rather than made explicit in the written text. Like Joan's "Our Best Thanksgiving," Ann's page keeps its secrets.

Conclusions: Listening in the Gaps

In the 1950s, Erving Goffman distinguished between backstage realms—those spaces in which individuals are free to formulate their "true" identities free from the gaze of audiences or outsiders—and frontstage regions (Goffman 1959, 112). Today, the line between front- and backstage is increasingly blurred as reality television, blogs, and myriad forms of real-time social media bring ever more intimate and mundane discourse into the lives of strangers. Yet at the turn of the millennium, before handheld technologies became ubiquitous and WYSIWYG (what-you-see-is-what-you-get) software enabled easy digital production, some women played up historical understandings of scrapbooks-as-praise-forms in order to focus attention on everyday tasks and relationships. Far from being introspective archives, these compiled books were exhibits that displayed everyday intimacies to a spectrum of publics; at the same time, the women I worked with took care to mediate reception, revealing information on their own terms. Though participation in what is now called "paper-based scrapbook-ing" has waned somewhat since its heyday in 2004 and 2005, these books and their digital counterparts continue to both push and mon-itor distinctions between public and private. Some books avoid writ-ten narratives, physically bury explicit texts in or behind other objects on the page, or include images and objects that merely index (rather than represent) painful or complex memories; to the casual observer, pages of this sort may come across as barren, hasty, superficial, or trite. But as I have demonstrated, these elisions and inclusions are no accident. And even when scrapbook creators do offer substantive written commentary, the meanings of their pages are always enriched and complicated during oral narration.

Another kind of talk is also important to understanding specific contexts of creation and display. While scrapbooks are certainly part of a postmodern move toward metacommentary in general, during the first decade of the twenty-first century scrapbooks and their makers negotiated public and private largely within organized "mediating structures," such as scrapbook groups in all their varieties, both virtual and face to face. Comprised of individuals from many social networks, these voluntary associations foster complex positioning with regard to public/private distinctions (cf. Evans 1997; Mechling 1989). At parties and in stores, convention centers, and online platforms, scrapbookers have cultivated spaces in which "rapport speech"—talk that takes pains to equalize status and functions as a display of caring— is appropriate and valued (Tannen 1990). Employing material and verbal strategies to cultivate interaction and supportive intimacy, scrapbooks-as-exhibitions can extend to public realms standards for evaluation and discourse that have long been associated with more private contexts.[36]

The recalibration of the public/private fractal discussed here is politically ambiguous. On the one hand, rapport management and strategic exclusion or undercoding reinforce conventional expectations about women's self-effacement and modes of managing aggression or desire (Maltz and Borker 1982; Simmons 2002; Tannen 1990). But they also rework the ideological and interactive norms associated with public spaces. In its broad outlines, this way of going public asks for a relativistic rather than hierarchical approach to evaluating others; it requires assigning value to diverse perspectives even while it demands the assertion of personal preferences. And it encourages the sharing of just those mundane details that make empathy across difference possible. As Jennifer González has suggested with regard to the public display of home altars, "To bring [this private practice] into the gallery . . . space is to transform a *rhetorical tactic* into a *rhetorical strategy*. It involves taking a stand in a public forum that is nevertheless grounded in an intimate and culturally specific practice" (González 1993, 90). Scrapbooks occupy similarly liminal spaces, offering representations important both for their "cult value"—their role in provoking personal contemplation—and for their "exhibition value"—the roles they play simply by being "on view" (Benjamin 1955 [1968], 226–27). These displays provide the social validation not readily available in "home mode" photography (Musello 1980); at the same time, they demonstrate how, in the words of folklorist Amy Shuman, the "unstable relations" of genres "become resources for sustaining the contradictions and discrepancies that are part of complex cultural situations" (Shuman 1993, 81).

This complex cultural situation involves ethnographers and museum educators, too: we work to make ideas and connections visible and accessible, to go public with research, to unearth ideas and options previously obscured by scholarly prose or languishing forgotten in archives and academic journals. Yet our engagements in the field also require careful listening and seeing—an awareness that our interlocutors are constantly calibrating access and disclosure, an acknowledgment that visibility has its risks as well as its benefits. Comprehensive and collaborative public humanities work thus requires attending to the specific ways genres and formal features are mobilized to claim, shape, and reconcile new spaces, audiences, and statuses—especially when those genres, and the people who create them, have not been considered noteworthy largely because we assume we already know them too intimately.

Acknowledgments

My deep thanks to all those I encountered during my fieldwork; your thoughtful reflections, good humor, and willingness to go public helped me see your craft and your labor in new, and I hope useful, ways. Thanks also to Indiana University (IU) for fellowship funding during my graduate studies and my fieldwork, and to colleagues at IU, The Ohio State University, The University of North Carolina at Chapel Hill, and Virginia Tech for offering comradeship and critique as I've revisited this work in recent years. Special gratitude goes to Dick Bauman, Sheila Bock, Katie Carmichael, Cassie Chambliss, Jason Jackson, Dorry Noyes, Kathy Roberts, Patricia Sawin, and Kyoim Yun for their brilliant support at many stages of thinking and writing, and to Gary Dunham and Janice Frisch at Indiana University Press for moving this project through with pleasant alacrity.

Notes

1. For example, Anderson and Goodenough (1929); Armstrong (2004); Buckler and Leeper (1991); Garvey (1996, 2003, 2013); Gear (1987); Greenhill (1981); Havens (2001); Helfand (2008); Katriel and Farrell (1991); McNeil (1968); Mecklenburg-Faenger (2007); Piepmeier (2009); Pigza (2016); Potter (1948, 1949); Rosenberg (1980); Stabile (2004); Theophano (2002); Titus (1976); Trimbur (2001); Tucker, Ott, and Buckler et al. (2006), Tucker (2013); and Zeitlin, Kotkin, and Baker et al. (1982).

2. Economic, technological, and life-cycle changes have since prompted some of these enthusiasts to turn to blogging, to designing and printing digital albums, or to other pursuits entirely. In 2006, more than seventeen million American households reportedly scrapbooked in some form, spending an estimated $2.5 billion on "memory craft" products in the process (Craft and Hobby Association 2008). By 2011, the projections were more modest: one scrapbook firm estimated that 4.5 million Americans scrapbooked as a hobby (about 1.7 percent of the American population, and 4.5 percent of women between the ages of 16 and 64; see Scrapbooking.com Magazine 2011), and by 2013, two major paper-based scrapbook retailers, Creative Memories and Archiver's, had declared bankruptcy and drastically altered operations. Archiver's, for instance, closed all of its forty–four brick-and-mortar stores in 2014, though it maintains an online presence (see Nancy Nally's reporting at www.scrapbookupdate.com).

3. See also Jon Kay, Daniel C. Swan, and James Cooley, this volume.

4. In part, this is due to the fact that scholars themselves have tended to study scrapbooks languishing in archives (e.g., Buckler and Leeper 1991; Helfand 2008; Langford 2001; Motz 1989; Tucker, Ott, and Buckler 2006). These lonely, decrepit albums have been parsed to glean important biographical data or to discover insights into broader social norms and conditions.

5. Scholars interested in genre politics examine the nexus of discourse that asserts not only what a particular genre "is" but also where it fits in regimes of value. Mikhail Bakhtin, among others, has demonstrated that the constellation of features that people assemble and name "as" genres change, as does the value attributed to their constituent parts. Attending to the politics of genre means asking which genres and genre features are dominant, authoritative, or trivialized, and why; it also means taking into account participant structures and conventional performance venues (e.g., Bakhtin 1981, 1986; Briggs and Bauman 1992; Gerhart 1992; Lawless 1997; Shuman 1993).

6. Because scrapbooks are unique objects heavy with idiosyncratic content, when encountered alone, divorced from their original contexts and makers, they require viewers to rely on convention in order to abstract the shared from the particular and simultaneously invest these abstractions with personal significance (cf. Abrahams 1969; Glassie 1972, 1989, 119; Stahl 1989). As in private

diaries—which are often "so terse they seem coded" (Bloom 1996, 25)—
scrapbooks are also rarely "self-coherent" but exhibit an alignment with other
texts and other performances by the very things they leave unsaid (Bloom 1996,
26). In the absence of the author, readers of these documents must tease out a
"mosaic portrait" (Bloom 1996, 27) of the book's creator based on intertextual
clues and their own experience. As Katherine Ott has observed about archived
scrapbooks, "The narrator who would have looked over the reader's shoulder and
excitedly pointed to the page, explaining it all, is long gone. So we are left with our
imagination and our desire to connect with the past" (Ott 2006, 40–41).

7. Though cultural demographics have since shifted in this regard, at the
height of "contemporary" paper-based scrapbooking in the early 2000s, the majority of self-identified enthusiasts were white mothers and grandmothers.
Thirty-nine percent of scrapbookers identified by the 2003 industry benchmark
survey had annual household incomes of less than $35,000; an additional 21
percent had annual household incomes of between $35,000 and $54,999, while
40 percent of scrapbooking households had annual incomes of $55,000 or high-
er (Primedia 2004, 4; see also Christensen 2011).

8. Transcripts from interviews in this chapter are presented using standard
techniques for ethnopoetic representation; line breaks and indented blocks, for
instance, highlight the rhythms of oral speech, signaling when interlocutors shift
into narration, quotation, interpretation, and the like. Note that an asterisk
indicates a smile in the voice, though not necessarily a full laugh; an open left
bracket represents overlapping speech, and italics signal emphasis. For a brief
introduction to the factors that motivate this approach to the presentation of
oral discourse, see Tedlock (1992).

9. Here and elsewhere, Joan signals her participation in (and as I will argue,
preference for) regional speechways and oral narrative traditions in several ways.
She uses a genitive (possessive) form to name the Kroger supermarket chain
("Krogers/Kroger's"; see Fruehwald 2012). More importantly, she uses the historical present/"narrative aspect" to mark the perspective she's taking with regard to the nested events being narrated, that is, the moment in the hospital and
the later construction of the page itself. When Joan reports, "So I says 'Our best
Thanksgiving' because Katie *made* it," *says* is in the present tense, suggesting that we
should envision the actual creation of the page as she speaks. (Not incidentally, *says*
also lexically conflates *speaking* and *writing/making*; she is quoting here the text that
she printed on the page.) This usage also takes a reflexive stance: as in the more
explicit "says I to myself," she is reflecting on her own actions (first-person pro-
noun) but in her role as narrator—so she uses a distanced third-person verb. For
an overview of the historical present tense and narrative aspect, see Vinogradov
2014; for other examples of "I says" in oral tradition, see Hymes 2003, 224–29.

10. Katherine Ott has noted that nineteenth- and early twentieth-century albums kept by physicians are also largely devoid of written commentary. However,
she concludes that in these books, which she characterizes as "organized reposi-
tories of knowledge" that focus on "public topics that reverberated in [the doctors'] personal lives" (Ott 2006, 38, 40), writing is unnecessary since unspoken
attitudes emerge from "transfixed text" such as newspaper clippings (37). In
books with more private or domestic foci, appropriate auxiliary texts are less

likely to be printed and more likely to be oral—or original expressive works like poetry that have been downloaded from the internet.

11. Drawing on the work of Alfred Schutz and Thomas Luckmann (1973), Richard Bauman has suggested that *entextualization*—the packaging of experience in bounded, transmissible discursive forms—facilitates the sharing, or socialization, of knowledge because it objectifies what would otherwise be ephemeral. Entextualization, he writes, "enhances the transferability, memorability, and repeatability of the encoded knowledge, now rendered as a text and thus as a durable unit of knowledge" (Bauman 2001, 109; Urban 1996).

12. Cf. Yankah's work on the oral elaboration of Akan royal staffs (Yankah 1995).

13. The fact that Joan could purchase patterned paper depicting the "owies" of childhood suggests that this kind of everyday documentation had become more prevalent in the twenty years since Chalfen's survey; certainly Facebook and other kinds of online blogging have dramatically changed this environment as well.

14. In her work on altar displays, Jennifer González has considered how arranged objects in general constitute "a vocabulary of persuasion, of memorial, of praise and display" (González 1993, 83; cf. Kirshenblatt-Gimblett 1989).

15. In 1992, near the end of high school, Sarah had followed her older sister's example and made a scrapbook about her teen years, but it was more than a decade before she briefly returned to the practice. Though she enjoys reading, quilting, antiquing furniture, and "organizing in general," she told me she usually avoids scrapbooking because "I picture something grand and glorious, and it's too expensive and time consuming to make." Instead, she places family pictures in a photo album (and later began a blog called "Sarah's Quilt"). Now that digitally produced photo books can be printed on demand, several of my field consultants mentioned that their husbands had become more involved in producing these family records.

16. At the time we spoke, Amy and Michelle each had several adolescent children; Amy was a past member of a writing club and managed the office of a local orthodontist, and Michelle was a project specialist at a medical laboratory.

17. Cf. Bauman (1986) and Bloom (1996). For an extended treatment, see Christensen (2009).

18. Compare Beverly Gordon, *Bazaars and Fair Ladies: The History of the American Fundraising Fair* (Gordon 1998), which discusses the dramatization and display of sewing, cooking, and decorating skills at bazaars, church suppers, and dramatic tea parties.

19. See www.craftster.org and www.sublimestitching.com for examples.

20. Top ten lists from www.bigpicturescrapbooking.com/teachers.php, accessed 2009; the site has since been revised. For more current bios, see http://www.bigpictureclasses.com/classes/see-all-teachers, especially the videos that accompany some of the profiles.

21. In some cultural contexts, expressions of modesty can actually key a performance frame; "I'm terrible at this, but . . ." can signal that someone is about to claim the stage (Bauman 1984 [1977]). It is possible that some scrapbook makers might declare an "unwillingness to assume responsibility" for effective display because they are cognizant of audience expectations but cannot meet them due to an inappropriate context, lack of necessary knowledge, or deficient skill

(Bauman 2004, 123), but among this population it is more likely that norms of modesty preclude such self-acknowledgment.

22. In the years after these discussions, Cherie launched her own brick-and–mortar gallery space where she taught art classes and displayed her own work; now a grandmother, she has closed the business in order to focus more on her own painting and multimedia work.

23. Renae painted a similar picture, noting that even when her children were in their twenties, "Whenever they had a new boyfriend or a girlfriend, the first thing they did was bring them home and they had to look through all *fifteen albums! [laughing] And everybody was excited! The friend was excited, and the kid was excited to remember it with em. That was amazing to me. Amazing that my boys, more often than my daughter, showed those albums to their girlfriends."

24. When *Chicago Tribune* reporter Pat Reardon interviewed Biegalski in 2004, she was a thirty-five-year-old resident of Lindenhurst, Illinois. A home day-care provider, her husband was a biologist at Abbott Labs.

25. "Ellen Scott" is a pseudonym.

26. "Renae Call" is a pseudonym.

27. For more on the ways imagined addressees affect the content, features, and style of an utterance, see Bakhtin (1986, 95–99).

28. "Kendra Barr" is a pseudonym.

29. Exploiting intertextuality—the ways texts or performances build upon, transform, and anticipate other texts (Bakhtin 1986; Bauman 2004; Genette 1982 [1997])—is one way to encourage viewers to create, contest, and reinterpret meanings. As anthropologists of ritual and festival have demonstrated, if a single material object encompasses multiple meanings, an *excess* of juxtaposed objects and symbols allows for playful realignments that encourage new interactions, dialogues, and ideas (Babcock 1978; La Fontaine 1985). Scrapbook compilations can generate the same interpretive variety; viewers must try to bear in mind "the original identity of the fragment or object and all of the history it brings with it; the new meaning it gains in association with other objects or elements; and the meaning it acquires as the result of its metamorphosis into a new entity" (Waldman 1992, 11).

30. Chalfen advises album researchers to look for "forces of suppression, deletion, alteration, or elimination" that counter or balance conventional "intensification, elaboration, exaggeration, and repetition" (Chalfen 1987, 92).

31. By 2009, Jane (a pseudonym) had added a third child to her family and was finishing a BFA in photography, in addition to working full-time as a professor in the social sciences.

32. Semiotician C. S. Peirce defined *icons* as those signs that demonstrate formal resemblance between the signifier and the signified (a map is an icon of a place, a photo is an icon of a person); *indexes* as signs that make meaning out of spatial or temporal contiguity to something else ("where there's smoke, there's fire"); and *symbols* as signs that have meaning due to cultural conventions (red means "stop" simply because people agree that it does; Peirce 1982, 53–56).

33. Stories elicited from senior citizens about their artistic productions (hooked wall hangings, memory paintings) are similarly nuanced and sometimes tragic; for example, Kay (2016) and Kirshenblatt-Gimblett and Kirshenblatt (2007).

34. In his discussion of phototherapy, Chalfen recognizes the potential of family photos to call up unpleasant memories (Chalfen 1987).

35. "Ann Russo" and "Carolyn" are pseudonyms. Ann left her job as a physician in order to care for her second child; she has since returned to work as a volunteer physician at a free clinic, and she also homeschools local teenagers, teaching everything from Latin to AP Government.

36. In their discussion of expressive forms historically produced by women, for instance, Melissa Meyer and Miriam Schapiro suggest that the intimate or local contexts in which handwork was seen and used ensured that it "was destined to be appreciated and admired" (Meyer and Schapiro 1978, 68).

References

Abel, Emily K. 2000. *Hearts of Wisdom: American Women Caring for Kin, 1850–1940.* Cambridge, MA: Harvard University Press.

Abrahams, Roger D. 1969. "The Complex Relations of Simple Forms." *Genre* 2: 104–28.

Anderson, John E., and Florence L. Goodenough. 1929. *The Modern Baby Book and Child Development Record.* New York: W. W. Norton and The Parents' Magazine.

Ang, Ien. 1985. *Watching Dallas: Soap Opera and the Melodramatic Imagination.* New York: Methuen.

Appadurai, Arjun. 1986. "Commodities and the Politics of Value." In *The Social Life of Things,* edited by Arjun Appadurai, 3–30. Cambridge: Cambridge University Press.

Armstrong, Erica R. 2004. "A Mental and Moral Feast: Reading, Writing, and Sentimentality in Black Philadelphia." *Journal of Women's History* 16: 78–102.

Babcock, Barbara A., ed. 1978. *The Reversible World.* Ithaca, NY: Cornell University Press.

Bakhtin, Mikhail. 1981. *The Dialogic Imagination.* Translated by Caryl Emerson and Michael Holquist. Austin: University of Texas Press.

———. 1986. *Speech Genres and Other Late Essays.* Translated by Vern W. McGee. Austin: University of Texas Press.

Barber, Karin. 2005. "Text and Performance in Africa." *Oral Tradition* 20: 264–77.

Baudrillard, Jean. 1968 (1996). *The System of Objects.* Translated by James Benedict. London: Verso.

Bauman, Richard. 1984 (1977). *Verbal Art as Performance.* Prospect Heights, IL: Waveland Press.

———. 1986. *Story, Performance, and Event: Contextual Studies of Oral Narrative.* Cambridge: Cambridge University Press.

———. 2001. "Mediational Performance, Traditionalization, and the Authorization of Discourse." In *Verbal Art across Cultures: The Aesthetics and Proto-Aesthetics of Communication,* edited by Hubert Knoblauch and Helga Kotthoff. Tübingen, Germany: Gunter Narr.

———. 2004. *A World of Others' Words: Cross-Cultural Perspectives on Intertextuality.* Oxford: Blackwell.

Benjamin, Walter. 1955 (1968). "The Work of Art in the Age of Mechanical Reproduction." In *Illuminations*, 219–53. New York: Harcourt, Brace & World.

Bloom, Lynn Z. 1996. "'I Write for Myself and Strangers': Private Diaries as Public Documents." In *Inscribing the Daily: Critical Essays on Women's Diaries*, edited by Suzanne L. Bunkers and Cynthia A. Huff, 23–37. Amherst: University of Massachusetts Press.

Briggs, Charles L., and Richard Bauman. 1992. "Genre, Intertextuality, and Social Power." *Journal of Linguistic Anthropology* 2: 131–72.

Buckler, Patricia P., and C. Kay Leeper. 1991. "An Antebellum Woman's Scrapbook as Autobiographical Composition." *Journal of American Culture* 14: 1–8.

Carl H. Pforzheimer Collection of Shelley and His Circle, The New York Public Library. 2016. "'Anna, may ev'ry Bliss be thine, . . .' (Felicia aged twelve to her aunt Anne Wagner)." New York Public Library Digital Collections. Accessed February 27. http://digitalcollections.nypl.org/items/510d47db-b637-a3d9-e040-e00a18064a99

Chalfen, Richard. 1987. *Snapshot Versions of Life*. Bowling Green, OH: Bowling Green State University Popular Press.

Challinor, Joan R., et al. 1979. "Family Photo Interpretation." In *Kin and Communities: Families in America*, edited by Allan J. Lichtman and Joan R. Challinor, 175–86. Washington, DC: Smithsonian Institution Press.

Christensen, Danille Elise. 1999. Field Notes on Scrapbooking in Indiana. Unpublished manuscript in the possession of the author.

———. 2003. Field Notes on Scrapbooking in Indiana. Unpublished manuscript in the possession of the author.

———. 2004a. Field Notes on Scrapbooking in North Carolina. Unpublished manuscript in the possession of the author.

———. 2004b. Field Notes on Scrapbooking in Ohio. Unpublished manuscript in the possession of the author.

_____. 2005a. Field Notes on Scrapbooking in Indiana. Unpublished manuscript in the possession of the author.

———. 2005b. Field Notes on Scrapbooking in Ohio. Unpublished manuscript in the possession of the author.

———. 2006. Field Notes on Scrapbooking in Utah. Unpublished manuscript in the possession of the author.

———. 2009. "Constructing Value: Women, Scrapbooking, and the Framing of Daily Experience." PhD diss., Indiana University.

———. 2011. "'Look at Us Now!': Scrapbooking, Regimes of Value, and the Risks of (Auto)ethnography." *Journal of American Folklore* 124: 175–210.

Clifford, James. 1994. "Collecting Ourselves." In *Interpreting Objects and Collections*, edited by Susan M. Pearce, 258–68. New York: Routledge.

Clifford, James, and George E. Marcus. 1986. *Writing Culture: The Poetics and Politics of Ethnography*. Berkeley: University of California Press.

Cowan-Bell, Tavel. 1998. "Putting Yourself Back in the Pictures." *Creating Keepsakes* September/October: 43–44.

Craft and Hobby Association. 2008. "National Scrapbook Day Media Alert." Press release downloaded from www.craftandhobby.org. 30 April. Accessed November 17.

de Certeau, Michel. 1984 (1988). *The Practice of Everyday Life.* Translated by Steven Rendall. Berkeley: University of California Press.

Dégh, Linda, and Andrew Vazsonyi. 1974. "The Memorate and the Proto-Memorate." *Journal of American Folklore* 87: 225–39.

Di Leonardo, Micaela. 1987. "The Female World of Cards and Holidays: Women, Families, and the Work of Kinship." *Signs: Journal of Women in Culture and Society* 12: 440–53.

Evans, Sara M. 1997. "Women's History and Political Theory: Toward a Feminist Approach to Public Life." In *Visible Women: New Essays on American Activism,* edited by Nancy A. Hewitt and Suzanne Lebsock, 119–39. Chicago: University of Illinois Press.

Fruehwald, Josef. 2012. "Have You Been in a Wawa's?" *Val Systems.* June 20. http://val-systems.blogspot.com/2012/06/have-you-been-in-wawas.html.

Gal, Susan. 2002. "A Semiotics of the Public/Private Distinction." *Differences: A Journal of Feminist Cultural Studies* 13: 77–95.

Garvey, Ellen Gruber. 1996. *The Adman in the Parlor: Magazines and the Gendering of Consumer Culture, 1880s to 1910s.* New York: Oxford University Press.

———. 2003. "Scissoring and Scrapbooks: Nineteenth-Century Reading, Remaking, and Recirculating." In *New Media: 1740–1915,* edited by Lisa Gitelman and Geoff Pingree, 206–27. Cambridge, MA: MIT Press.

———. 2013. *Writing with Scissors: American Scrapbooks from the Civil War to the Harlem Renaissance.* New York: Oxford University Press.

Gear, Josephine. 1987. "The Baby's Picture: Woman as Image Maker in Small-Town America." *Feminist Studies* 13: 419–42.

Gelber, Steven M. 1999. *Hobbies: Leisure and the Culture of Work in America.* New York: Columbia University Press.

Genette, Gérard. 1982 (1997). *Palimpsests: Literature in the Second Degree.* Translated by Channa Newman and Claude Doubinsky. Lincoln: University of Nebraska Press.

Gerhart, Mary. 1992. *Genre Choices, Gender Questions.* Norman: University of Oklahoma Press.

Glassie, Henry. 1972. "Folk Art." In *Folklore and Folklife: An Introduction,* edited by Richard M. Dorson, 253–80. Chicago: University of Chicago Press.

———. 1982. *Passing the Time in Ballymenone: Culture and History of an Ulster Community.* Bloomington: Indiana University Press.

———. 1989. *The Spirit of Folk Art: The Girard Collection at the Museum of International Folk Art.* New York: Harry N. Abrams.

———. 2000. *Vernacular Architecture.* Bloomington: Indiana University Press.

Goffman, Erving. 1959. *The Presentation of Self in Everyday Life,* Revised edition. New York: Anchor.

———. 1961. *Encounters: Two Studies in the Sociology of Interaction.* Indianapolis, IN: Bobbs-Merrill.

González, Jennifer A. 1993. "Rhetoric of the Object: Material Memory and the Artwork of Amalia Mesa-Bains." *Visual Anthropology Review* 9: 82–91.

Goodman, Nelson. 1968. *Languages of Art: An Approach to a Theory of Symbols.* Indianapolis, IN: Bobbs-Merrill.
Goodwin, Marjorie Harness. 1980a. "Directive-Response Speech Sequences in Girls' and Boys' Task Activities." In *Women and Language in Literature and Society,* edited by Sally McConnell-Ginet, Ruth Borker, and Nelly Furman, 157–73, New York: Praeger.
———. 1980b. "He-Said-She-Said: Formal Cultural Procedures for the Construction of a Gossip Dispute Activity." *American Ethnologist* 7: 674–95.
Gordon, Beverly. 1998. *Bazaars and Fair Ladies: The History of the American Fundraising Fair.* Knoxville: University of Tennessee Press.
Greenhill, Pauline. 1981. *So We Can Remember: Showing Family Photographs,* National Museum of Man Mercury Series, Canadian Centre for Folk Culture Studies, Paper No. 36, Ottawa, National Museums of Canada.
Havens, Earle. 2001. *Commonplace Books: A History of Manuscripts and Printed Books from Antiquity to the Twentieth Century.* New Haven, CT: Beinecke Rare Book and Manuscript Library.
Helfand, Jessica. 2008. *Scrapbooks: An American History.* New Haven, CT: Yale University Press.
Honko, Lauri. 1964. "Memorates and the Study of Folk Beliefs." *Journal of the Folklore Institute* 1: 5–19.
Hymes, Dell. 1975. "Breakthrough into Performance." In *Folklore: Performance and Communication,* edited by Dan Ben-Amos and Kenneth S. Goldstein, 11–74. The Hague, the Netherlands: Mouton.
———. 2003. *Ethnography, Linguistics, Narrative Inequality: Toward an Understanding of Voice.* London: Taylor & Francis Group.
Jakobson, Roman. 1960. "Closing Statement: Linguistics and Poetics." In *Style in Language,* edited by Thomas A. Sebeok, 350–77. Cambridge, MA: MIT Press.
Jones, Michael Owen. 1995. "Why Make (Folk) Art?" *Western Folklore* 54: 253–76.
Julian, Stacy, and Terina Darcey. 1997. *Core Composition: Step-by-Step Layout Design for your Scrapbooks.* Salt Lake City, UT: Apple of Your Eye.
Kalčik, Susan. 1975. "'. . . Like Ann's Gynecologist or the Time I Was Almost Raped': Personal Narratives in Women's Rap Groups." In *Women and Folklore,* edited by Claire R. Farrer, 3–11. American Folklore Society Bibliographical and Special Series 28. Austin: University of Texas Press.
Katriel, Tamar, and Gerry Philipsen. 1981. "'What We Need Is Communication': 'Communication' as a Cultural Category in Some American Speech." *Communication Monographs* 48: 301–17.
Katriel, Tamar, and Thomas Farrell. 1991. "Scrapbooks as Cultural Texts: An American Art of Memory." *Text and Performance Quarterly* 11: 1–17.
Kay, Jon. 2016. *Folk Art and Aging: Life-Story Objects and Their Makers.* Bloomington: Indiana University Press.
Kirshenblatt-Gimblett, Barbara. 1989. "Objects of Memory: Material Culture as Life Review." In *Folk Groups and Folklore Genres: A Reader,* edited by Elliott Oring, 329–38. Logan: Utah State University Press.
———. 1995. "Theorizing Heritage." *Ethnomusicology* 39: 367–80.
———. 1998. "Secrets of Encounter." In *Destination Culture: Tourism, Museums, and Heritage,* 249–56. Berkeley: University of California Press.

————. 2006. "World Heritage and Cultural Economics." In *Museum Frictions: Public Cultures/Global Transformations*, edited by Ivan Karp, Corinne A. Kratz, Lynn Szwaja, and Tomás Ybarra-Frausto, with Gustavo Buntinx, Barbara Kirshenblatt-Gimblett, and Ciraj Rassool, 161–202. Durham, NC: Duke University Press.

Kirshenblatt-Gimblett, Barbara, and Mayer Kirshenblatt. 2007. *They Called Me Mayer July.* Berkeley: University of California Press.

La Fontaine, Jean. 1985. *Initiation.* Harmondsworth, England: Penguin.

Langellier, Kristin M. 1992. "Show and Tell in Contemporary Quiltmaking Culture." *Uncoverings: Research Papers of the American Quilt Study Group* 13: 127–47.

Langford, Martha. 2001. *Suspended Conversations: The Afterlife of Memory in Photographic Albums.* Montreal: McGill-Queen's University Press.

Lawless, Elaine J. 1997. "Women's Folk and Popular Arts: The Need for a Grounded Theory." In *The Material Culture of Gender, the Gender of Material Culture*, edited by Katharine Martinez and Kenneth L. Ames, 197–216. Hanover, NH: University Press of New England.

"Local Woman's Life Looks Bearable in Scrapbook." 2004. *The Onion: America's Finest News Source.* December 8. www.theonion.com.

Malcolm, Janet. 1976. "Diana and Nikon." *New Yorker,* April 26: 133–38.

Maltz, Daniel N., and Ruth A. Borker. 1982. "A Cultural Approach to Male–Female Miscommunication." In *Language and Social Identity,* edited by John J. Gumperz, 196–216. Cambridge: Cambridge University Press.

McDermott, Sinead. 2002. "Memory, Nostalgia, and Gender in a Thousand Acres." *Signs: Journal of Women in Culture and Society* 28: 389–407.

McLeod, Norma, and Marcia Herndon. 1975. "The Bormliza: Maltese Folksong Style and Women." In *Women and Folklore: Images and Genres,* edited by Claire R. Farrer, 81–100. Prospect Heights, IL: Waveland Press.

McNeil, W. K. 1968. "The Autograph Album Custom: A Tradition and Its Scholarly Treatment." *Keystone Folklore Quarterly* 13: 29–40.

Mechling, Jay. 1989. "Mediating Structures and the Significance of University Folk." In *Folk Groups and Folklore Genres: A Reader,* edited by Elliott Oring, 339–49. Logan: Utah State University Press.

Mecklenburg-Faenger, Amy L. 2007. "Scissors, Paste and Social Change: The Rhetoric of Scrapbooks of Women's Organizations, 1875–1930." PhD diss., The Ohio State University.

Meyer, Madonna Harrington, ed. 2000. *Care Work: Gender, Class, and the Welfare State.* New York: Routledge.

Meyer, Melissa, and Miriam Schapiro. 1978. "Waste Not, Want Not: An Inquiry into What Women Saved and Assembled." *Heresies* 4: 66–69.

Motz, Marilyn F. 1989. "Visual Autobiography: Photograph Albums of Turn-of-the-Century Midwestern Women." *American Quarterly* 41: 63–92.

Musello, Christopher. 1980. "Studying the Home Mode: An Exploration of Family Photography and Visual Communication." *Studies in Visual Communication* 6: 23–42.

Noyes, Dorothy. 1995. "Group." *Journal of American Folklore* 108: 449–78.

Ochs, Elinor. 1992. "Indexing Gender." In *Rethinking Context: Language as an Interactive Phenomenon*, edited by Alessandro Duranti and Charles Goodwin, 335–58. New York: Cambridge University Press.

Ott, Katherine. 2006. "Between Person and Profession: The Scrapbooks of Nineteenth-Century Medical Practitioners." In *The Scrapbook in American Life*, edited by Susan Tucker, Katherine Ott, and Patricia P. Buckler, 29–41. Philadelphia: Temple University Press.

Peirce, Charles S. 1982. *The Writings of Charles S. Peirce: A Chronological Edition*, vol. 2. Bloomington: Indiana University Press.

Piepmeier, Alison. 2009. *Girl Zines: Making Media, Doing Feminism.* New York: New York University Press.

Pigza, Jessica. 2016. "Romantic Remix: Ladies' 'Friendship' Albums." In *Frankenstein: The Afterlife of Shelley's Circle.* Columbus, OH: Biblion; New York Public Library. Accessed February 27. http://exhibitions.nypl.org/biblion/node/2738.

Pocius, Gerald L. 1991. *A Place to Belong: Community Order and Everyday Space in Calvert, Newfoundland.* Athens: University of Georgia Press.

Potter, Charles Francis. 1948. "Round Went the Album." *New York Folklore Quarterly* 4: 4–14.

———. 1949. "Autograph Album Rimes." In *Wagnall's Standard Dictionary of Folklore, Mythology, and Legend*, edited by Maria Leach, 94–96. New York: Funk & Wagnalls.

Primedia. 2004. *The National Survey of Scrapbooking in America: A Benchmark Study of the Scrapbooking Industry Sponsored by* Creating Keepsakes *Magazine and* Craftrends *Magazine.* Bluffdale, UT: Primedia.

Pye, David. 1968 (1978). *The Nature and Art of Workmanship.* Cambridge: Cambridge University Press.

Radner, Joan N., and Susan S. Lanser. 1987. "The Feminist Voice: Strategies of Coding in Folklore and Literature." *Journal of American Folklore* 100: 412–25.

Radway, Janice A. 1984 (1991). *Reading the Romance: Women, Patriarchy, and Popular Literature.* Chapel Hill: University of North Carolina Press.

Reardon, Patrick T. 2004. "The Pages of My Life, by Me." *Chicago Tribune* March 18: N1.

Rosenberg, Neil V. 1980. "'It Was a Kind of Hobby': A Manuscript Song Book and Its Place in Tradition." In *Folklore Studies in Honour of Herbert Halpert*, edited by Kenneth S. Goldstein and Neil V. Rosenberg, 315–34. St. John's, NL: Memorial University of Newfoundland Folklore and Language Archive.

Rueger, Lydia. 2001. "Just a Word About Me," *Memory Makers* Jan/Feb: 58–63.

Sawin, Patricia E. 1992. "'Right Here Is a Good Christian Lady': Reported Speech in Personal Narratives." *Text and Performance Quarterly* 12: 193–211.

———. 2002. "Performance at the Nexus of Gender, Power, and Desire: Reconsidering Bauman's *Verbal Art* from the Perspective of Gendered Subjectivity as Performance." *Journal of American Folklore* 115: 28–61.

Schutz, Alfred, and Thomas Luckmann. 1973. *The Structures of the Life-World*, vol. 1. Edited by Richard M. Zaner and H. Tristram Engelhardt Jr. Evanston, IL: Northwestern University Press.

Scott, James C. 1985. *Weapons of the Weak: Everyday Forms of Peasant Resistance*. New Haven, CT: Yale University Press.

Scrapbooking.com Magazine. 2011. "Scrapbooking.com 2011 Magazine Profile." Accessed December 5, 2011. http://scrapbooking.com/sales/files/Magazine Profile.pdf.

Shuman, Amy. 1993. "Gender and Genre." In *Feminist Theory and the Study of Folklore*, edited by Susan Tower Hollis, Linda Pershing, and M. Jane Young, 71–88. Urbana: University of Illinois Press.

Simmons, Rachel. 2002. *Odd Girl Out: The Hidden Culture of Aggression in Girls*. New York: Harcourt.

Smith, Sidonie. 1987. *A Poetics of Women's Autobiography: Marginality and the Fictions of Self-Representation*. Bloomington, IN: Indiana University Press.

Spence, Jo, and Patricia Holland, eds. 1991. *Family Snaps: The Meanings of Domestic Photography*. London: Virago.

Stabile, Susan. 2004. *Memory's Daughters: The Material Culture of Remembrance in Eighteenth-Century America*. Ithaca, NY: Cornell University Press.

Stahl, Sandra K. D. 1989. *Literary Folkloristics and the Personal Narrative*. Bloomington: Indiana University Press.

Stewart, Kathleen. 1996. *A Space on the Side of the Road: Cultural Poetics in an "Other" America*. Princeton, NJ: Princeton University Press.

Tannen, Deborah. 1985. "Relative Focus on Involvement in Oral and Written Discourse." In *Literacy, Language, and Learning: The Nature and Consequences of Reading and Writing*, edited by David R. Olson, Nancy Torrance, and Angela Hildyard, 124–47. Cambridge: Cambridge University Press.

———. 1990. *You Just Don't Understand: Women and Men in Conversation*. New York: William Morrow.

Tedlock, Dennis. 1992. "Ethnopoetics." In *Folklore, Cultural Performances, and Popular Entertainments: A Communications-centered Handbook*, edited by Richard Bauman, 81–85. New York: Oxford University Press.

Theophano, Janet. 2002. *Eat My Words: Reading Women's Lives through the Cookbooks They Wrote*. New York: Palgrave.

Titus, Sandra L. 1976. "Family Photographs and Transition to Parenthood." *Journal of Marriage and Family* 38: 525–30.

Trimbur, John. 2001. "Popular Literacy: Caught between Art and Crime." In *Popular Literacy: Studies in Cultural Practices and Poetics*, edited by John Trimbur, 1–16. Pittsburgh, PA: University of Pittsburgh Press.

Tucker, Susan. 2013. "Tacitly the Work of Women: Personal Archives and the Public Memory of Families." In *Perspectives on Women's Archives*, edited by Tanya Zanish-Belcher, 147–76. Chicago, IL: Society of American Archivists.

Tucker, Susan, Katherine Ott, and Patricia P. Buckler, eds. 2006. *The Scrapbook in American Life*. Philadelphia, PA: Temple University Press.

Ulrich, Laurel T. 2001. *The Age of Homespun: Objects and Stories in the Creation of an American Myth*. New York: Random House.

Urban, Greg. 1996. "Entextualization, Replication, and Power." In *Natural Histories of Discourse*, edited by Michael Silverstein and Greg Urban, 21–44. Chicago: University of Chicago Press.

Vinogradov, Igor. 2014. "Aspect Switching in Tzotzil (Mayan) Narratives." *Oklahoma Working Papers in Indigenous Languages* 1: 39–54.

Waldman, Diane. 1992. *Collage, Assemblage, and the Found Object.* London: Phaidon Press.

Wallace, James D. 1990. "Hawthorne and the Scribbling Women Reconsidered." *American Literature* 62: 201–22.

Weiner, Annette B. 1992. *Inalienable Possessions: The Paradox of Keeping-While-Giving.* Berkeley: University of California Press.

Wilson, William A. 1995. "Mormon Narratives: The Lore of Faith." *Western Folklore* 54: 303–26.

Woolf, Virginia. 1929 (1989). *A Room of One's Own.* New York: Harcourt, Brace and Co.

Yankah, Kwesi. 1994. "Visual Icons and the Akan Concept of Proverb Authorship." *Passages: A Chronicle of the Humanities* 7: 1–3.

———. 1995. *Speaking for the Chief: Okyeame and the Politics of Akan Royal Oratory.* Bloomington: Indiana University Press.

Zeitlin, Steven J., Amy J. Kotkin, and Holly Cutting Baker. 1982. *A Celebration of American Family Folklore: Tales and Traditions from the Smithsonian Collection.* New York: Pantheon Books.

DANILLE ELISE CHRISTENSEN is Assistant Professor of Public Humanities in the Department of Religion and Culture at the Virginia Polytechnic Institute and State University (Virginia Tech).

Michael Paul Jordan

3 Depictions of Women and Warfare in Kiowa Drawings from Fort Marion: Reassessing Nineteenth-Century Kiowa Gender Roles

ANTHROPOLOGIST BEATRICE MEDICINE notes that the androcentric bias of early ethnographers of the Plains Indian societies led them to largely ignore the variety of socially sanctioned female roles, writing that "unfortunately," their reliance on "such notions as 'warrior society,' 'male dominant,' and other male supremacist expressions have set the tone for the analysis of male and female behavior in Plains Indian society" (Medicine 1983, 276). Medicine argues that assumptions regarding the rigidity of gender norms prevented nine-teenth-century and early twentieth-century anthropologists from exploring female participation in masculine pursuits. She laments that, "Regrettably, the kinds of data that might illuminate and clarify the nature of female role reversals in Plains Indian societies have not been recorded," adding that, "it is now too late to recover such material with any degree of depth" (Medicine 1983, 276–77). Fortunately, the Kiowa case marks an important exception. Nineteenth-century Kiowa drawings document female participation in warfare, and the testimony of Kiowa elders interviewed during the 1935 Santa Fe Laboratory of Anthropology field school (LaBarre 1935a, 1935b) provide valuable insights into the motivations underlying female participation in such endeavors.

Kiowa drawings produced between 1875 and 1878 at Fort Marion in St. Augustine, Florida, and interviews with Kiowa elders conducted by members of the 1935 Santa Fe Laboratory of Anthropology field school document nineteenth-century Kiowa women's participation in martial pursuits, challenging the prevailing tendency to view

Plains Indian warfare as an exclusively male domain (Ewers 1994, 325). Furthermore, these sources demonstrate that Kiowa women made contributions to the success of raiding parties that extended beyond performing camp chores such as cooking. Taken together, the drawings and interviews reveal that Kiowa women actively engaged in horse stealing and combat.

Nineteenth-Century Plains Indian Pictographic Art

Nineteenth-century Plains Indian pictographic art is widely recognized as having been a male domain (Keyser, Sundstrom, and Poetschat 2006, 53; Petersen 1971, 15–16; Szabo 1994, 3). As Candace Greene notes, "Men were the producers of Plains graphic art as well as the primary audience for which it was intended" (Greene 2001, 209). Men created autobiographical drawings that celebrated their martial achievements, as well as biographical drawings that chronicled their close comrades' exploits (Berlo 1990, 133; Harris 1989, 10; Keyser, Sundstrom, and Poetschat 2006, 53; Petersen 1971, 17; Szabo 1994, 10, 15; 1984, 368; Wong 1989, 297). Discussing these drawings, Joyce Szabo notes that, "Battles and horse raids are, by far, the most prevalent of all subjects depicted" (Szabo 1984, 368). Many of the scenes consisted of a mounted protagonist locked in combat with one or more foes (Szabo 1984, 370). Men painted their exploits on bison robes, which when worn served as a chronicle of the owner's war honors. In the latter half of the nineteenth century, men created drawings on paper, often employing bound books (Greene 2004, 21–22; Petersen 1971, 21–23, 25–26; Szabo 1984, 368; 1994, 10, 15).

In nineteenth-century Plains Indian graphic art, depictions of women typically fall into one of two categories: images of intertribal warfare or courting scenes. Male artists frequently depicted women as victims of male violence and objects of male sexual desire (Ewers 1994, 325–27; Greene 2001, 209). As John Ewers points out in his essay "Women's Roles in Plains Indian Warfare," the killing of enemy women was an established part of intertribal warfare in the nineteenth century. There was no stigma attached to the act and warriors celebrated their achievements in this arena, creating drawings in which they rendered themselves dispatching enemy women (Ewers 1994, 325–27).[1]

In other drawings (Fig. 3.1), men depict themselves counting coup on enemy women (Berlo 1996, 94–95, 108–109, 182–83, 186–87; Keyser, Sundstrom, and Poetschat 2006, 52; Petersen 1990, LXXII–LXXIII). Counting coup, striking an enemy without inflicting a mortal wound, was regarded as a deed worthy of recognition (Grinnell 1910, 297–300; Smith 1938, 427). Among the Cheyenne, counting coup on a woman from an enemy tribe carried as much prestige as counting coup on an enemy warrior (Grinnell 1910, 296–97, 302). In the 1890s, George Bird Grinnell interviewed Southern Cheyenne men who had participated in the intertribal warfare of the mid-nineteenth century. The warriors recounted their exploits for Grinnell, who recorded them in a small notebook now in the collection of the National Anthropological Archives. The notebook's contents reveal that several of the men, including Wolf Face and Little Chief, had counted coup on enemy women (Grinnell 1897).

FIGURE 3.1.
A Cheyenne warrior counts coup on two enemy women using a cavalry saber. The action in this scene moves from right to left. The mounted warrior first struck the woman in the red-striped blouse and skirt before riding on and striking the second woman. Unidentified Cheyenne Artist, ca. 1879–1885, Black Horse Ledger, plate 45. Mandeville Library and Plains Indian Ledger Art Publishing Project, University of California, San Diego, La Jolla, California. View the complete book at plainsledgerart.org.

Courting scenes highlight encounters between male suitors and women. Kiowa drawings (Fig. 3.2) frequently depict the courting couple wrapped in a blanket or conversing under the watchful eye of a female chaperone. Other courting scenes include details that suggest encounters of a more illicit nature. Greene notes that an artist might include a bucket or an axe in a drawing to signify that he intercepted a woman while she was fetching water or gathering firewood. The implication is that the encounter was unchaperoned. Nevertheless, courting scenes are typically rather staid, with male sexual conquests implied rather than explicitly depicted (Greene 2001, 209–10). While Plains Indian drawings that depict couples engaged in sexual intercourse are extremely rare, they do exist (Calloway 2012, plate 103). James D. Keyser, Linea Sundstrom, and George Poetschat (2006, 52) note that courting scenes typically depict women as the passive objects of male attention. For example, in the vast majority of nineteenth-century Cheyenne drawings, women occupy the same structural position as game animals and enemy

FIGURE 3.2.
A man, wrapped in a bison robe, courts a young woman within the confines of camp. The woman's chaperone, who wears a child in a lattice cradle on her back, observes the couple's interactions from a short distance away. Unidentified Kiowa artist, ca. 1875–1878, MS 98-54. National Anthropological Archives, Smithsonian Institution, Washington, DC.

warriors. In the drawings, the action moves from right to left. The male protagonists appear on the right while their intended quarry, be it an enemy warrior, a bison, or a woman, appears on the left. Based on the structural similarities between these drawings, Greene (1985, 1996a) concluded that Cheyenne men viewed courting as analogous to warfare and hunting. The Kiowa also recognized sexual conquests as male exploits deserving of recognition and a man's escapades bolstered his standing among his peers (LaBarre 1935a, 545; 1935b, 81, 93; Mishkin 1940 [1992], 37; Parsons 1929, 91). However, a subset of Kiowa drawings from Fort Marion presents women in an entirely different light, depicting them neither as the victims of male violence or the objects of male desire. These drawings document women's participation in the arena of warfare.

The Kiowa Pictographic Art Tradition

The drawings produced at Fort Marion are part of a larger Kiowa tradition of graphic art. While there is only one extant example of Kiowa pictographic art produced before Fort Marion, there is ample evidence that Kiowa men produced depictions of their martial exploits during the pre-reservation period. In 1845, the Cheyenne chief Sleeping Bear gave the rights to a painted tipi design to the Kiowa chief Little Bluff—a gesture intended to commemorate a peace treaty that the two tribes had entered into five years earlier. The bilaterally asymmetrical tipi design employed a series of stripes on the south side and depictions of battle scenes on the north side. Known as the *Tipi with Battle Pictures*, it was the only Kiowa tipi design that featured combat scenes. When the painted tipi cover became faded or worn, the design was renewed and a new cover was painted. At that time, the owner of the design would select the warriors whose martial exploits he wished to feature on the tipi cover. In Kiowa society, war honors were regarded as a form of intellectual property, and a man's deeds could not be depicted without his permission. A warrior might paint his own exploits or narrate his deeds while a more talented artist rendered them (Ewers 1978, 15–16; Greene 1993, 70–72; Greene and Drescher 1994, 422–23; Jordan and Swan 2011, 197–98).

Additional evidence that there was an established Kiowa pictographic art tradition in the pre-reservation period comes from the writings of De Benneville Randolph Keim, a reporter who accompanied General Sheridan's 1868 campaign on the Southern Plains. In December 1868, the US military seized the Kiowa leaders Lone Wolf and White Bear and held them prisoner in an effort to compel the Kiowa bands to report to Fort Cobb (Nye 1968, 138–39). During their brief confinement, the men depicted their martial exploits on a bison robe, which they sold to Keim. Through the medium of the painted hide, Lone Wolf and White Bear sought to convey information to their captors regarding their war records and social standing. While the drawings themselves have not survived, Keim's account suggests that Kiowa warriors were accustomed to depicting their war honors (Harris 1989, 11; Keim 1885, 223–24).

The only extant example of pre-reservation Kiowa pictographic art is a painted bison calf hide. Dr. Edward Palmer collected the hide in 1868 while serving as the government physician assigned to the Kiowa–Comanche Agency. The painting depicts a dismounted warrior, wearing a buffalo horn bonnet and wielding a bow. A dashed line marks his retreat from his wounded horse. The warrior is surrounded by a party of armed Whites. Four of these men are depicted wearing civilian attire. The remaining members of the party are represented by detailed depictions of firearms (Greene 1997, 44–45). The Kiowa celebrated the achievements of men who had survived being surrounded by enemy forces, and this hide painting likely commemorates such an event (Jordan 2012, 24; Jordan and Swan 2011, 198).

Prior to Fort Marion, Kiowa pictographic art focused on the depiction of male martial exploits. It remains unclear when Kiowa artists first adopted paper as a drawing medium. The one extant example of Kiowa pictographic art that predates Fort Marion is executed on a bison calf hide, and the accounts referenced above likewise describe works on hide. While the Cheyenne were certainly producing drawings on paper during the 1860s, scholars have not identified any Kiowa works on paper from this period (Petersen 1971, 25–26).

Kiowa Drawings from Fort Marion

Few Plains Indian drawings on paper have attracted more attention than those created by Kiowa, Cheyenne, and Arapaho men imprisoned at Fort Marion in St. Augustine, Florida. Following the end of the Red River War in 1875, the US military exiled seventy-two individuals from the Southern Plains tribes to Florida. Among the prisoners were twenty-seven Kiowa men. Over the course of their three-year incarceration, a number of the prisoners created drawings on paper (Szabo 2012, 233–34).

Several factors have been cited to explain the production of art at Fort Marion. Scholars have argued that the prisoners were motivated by feelings of nostalgia. According to this line of reasoning, the artists' depictions of life on the Plains prior to their incarceration and exile to Florida reflect their deep-seated feelings of homesickness (Berlo 1982, 11; Szabo 1984, 370; 1994, 72–73; 2007, 41–42; 2011, 31). The existence of a commercial market for the prisoner's art has also been cited as a motivating factor in the production of drawings. During the 1870s, St. Augustine was a popular vacation spot, and tourists frequently visited Fort Marion to catch a glimpse of the warriors. For these curious individuals, the drawings served as a memento of their encounter with a cultural other (Szabo 2007, 25–27; 2011, 20; Viola 1998, 17). In addition, Lieutenant Richard Henry Pratt, the officer in charge of the prisoners, encouraged the men to draw and provided them with art supplies. Pratt, an ardent proponent of assimilation, occasionally presented the prisoners' drawings to patrons whom he felt were well positioned to support his Indian education agenda (Szabo 2001, 50–51).

Scholars who have studied the drawings produced at Fort Marion have emphasized the difference between these works and earlier drawings produced by Plains Indian artists (Harris 1989, 11–12; Szabo 1984). Several of the Fort Marion artists documented their journey from Fort Sill to Florida and their experiences as prisoners. Others recorded views of Fort Marion, as well as the city of St. Augustine and its environs (Berlo 1990, 135; Harris 1989, 12). The men also drew scenes of their life on the plains. While depictions of martial exploits predominated in pre-reservation Plains Indian graphic art, Fort Marion artists more frequently drew scenes of hunting and courting. They also explored new subject matter, including warrior society ceremonials and religious rituals (Harris 1989, 12; Szabo 1984, 367–70). Szabo notes that, "Although occasional battle images did appear in the drawing books, most of the works were peaceful, nostalgic illustrations of other aspects of the traditional Plains life-style" (Szabo 1984, 367).

Bird Chief, White Horse, and Koba

Among the Kiowa drawings from Fort Marion that embrace martial themes are seven images that depict women's participation in warfare. The drawings were created by three men: Bird Chief, White Horse, and Koba.[2] Biographical information on Bird Chief is scant. He was known by at least two other names Bad Eye and Bird Medicine. At the time of his incarceration in 1875, he was forty-three years old. The roster of prisoners compiled by Pratt identifies him as a "warrior and leader" but not a chief (Pratt 2003, 141). This suggests that Bird Chief was not a residence band leader or *topatoki*. However, he had distinguished himself as a successful *toyopki* or war expedition leader (Petersen 1971, 208). Two of the raids that Bird Chief organized are well documented. In the summer of 1872, he led a raid into Kansas. His party was successful, capturing a number of mules (Mooney 1979, 335–36). He led another raid in the winter of 1873 and on November 21, members of his party killed a freighter who was traveling between the North Fork of the Canadian River and Camp Supply. His participation in this last incident was among the reasons listed for his exile at Fort Marion (Petersen 1971, 208; Pratt 2003, 141).

On August 22, 1874, Bird Chief was present at the Wichita Agency during the unrest that precipitated the Red River War. Having subsequently fled the reservation, he participated in the siege of Captain Willis Lyman's supply train (Pratt 2003, 141). Following this engagement, Bird Chief opted not to accompany the Kiowa faction that sought refuge in the Llano Estacado region of Texas, choosing instead to surrender. On October 3, 1874, he surrendered at the Cheyenne and Arapaho Agency at Darlington along with 150 others. He was eventually relocated to Fort Sill, where he was confined until he was transported to Fort Marion in St. Augustine, Florida (Petersen 1971, 208).

Following his release in 1878, Bird Chief returned to the Kiowa Reservation (Lookingbill 2006, 164, 166). He survived until at least 1892. Bird Chief fell ill in the spring or summer of that year and was attended to by several Kiowa doctors who possessed medicine or spiritual power (Mooney 1979, 377). Whether he recovered from his illness is not recorded. Regardless, Bird Chief's choice of treatment demonstrates that he retained his Kiowa religious beliefs late into his life.

White Horse (Fig. 3.3) was younger than Bird Chief when he was incarcerated; records indicate that he was born in 1847. Despite his relative youth, he had secured quite a reputation for himself. Out of the twenty-seven Kiowa prisoners, he was regarded as one of the four most prominent. In addition to being recognized as a war leader or *toyapki*, White Horse had gained a following and established himself as a *topatoki* or residence band leader (Petersen 1971, 11–112, 119). He was also a whipman or officer in the *Adoltoyui* or Young Mountain Sheep Society, one of the Kiowa warrior societies (Parsons 1929, 92–93; Petersen 1971, 112).

Unlike White Horse and Bird Chief, Koba (Wild Horse) had not risen to a position of prominence in Kiowa society. He was neither a recognized war leader nor the head of a residence band. Born in 1846, he was approximately twenty-six years old when he was incarcerated.

FIGURE 3.3.
This photograph of White Horse was made five years before his incarceration at Fort Marion. In the portrait, he wears a short hairpipe breastplate of a style popular on the Southern Plains. William S. Soule, 1870, NAA INV 01625402. National Anthropological Archives, Smithsonian Institution, Washington, DC.

He had participated in horse raids in Mexico and Texas, including an 1872 raid led by Man Who Walks above the Ground that targeted settlements along the Brazos River. Like Bird Chief, Koba was at the Wichita Agency on August 22, 1874, when fighting broke out.

Fort Marion Sketchbooks as Objects of Material Culture

The Kiowa drawings discussed in this essay are not only valuable sources of ethnographic data but objects of material culture in their own right. Each of the drawings was part of a larger bound sketch-book. One striking difference between the bound books employed by the Fort Marion artists and those utilized by Plains warriors in the pre-reservation period concerns their origins. Castle McLaughlin notes that almost all of the pre-reservation Cheyenne and Lakota drawings on paper are executed on the pages of books or documents that were captured from Whites. These books functioned as war trophies, material reminders of past victories. McLaughlin proposes the term "war books" to distinguish these objects from books that lack this association with warfare. She explains that, "a war book is significant as an object: a Euro-American document captured in battle that has been modified by Plains warriors, primarily through the addition and overlay of drawings, in order to appropriate the power of their ene-mies" (McLaughlin 2013, 6). While some Kiowa drawings from Fort Marion depict martial themes, the books in which these drawings appear were not war trophies.

 While a few Fort Marion drawings were executed on ruled paper or the pages of autograph books, most were placed in commercially produced sketchbooks. Lieutenant Pratt, the officer in charge of the prisoners, encouraged their artistic production and furnished them with materials. In September 1876, he ordered two dozen drawing books for the prisoners' use (Petersen 1971, 70). Occasionally, pri-vate parties commissioned drawings from the prisoners and furnished them with sketchbooks (Szabo 2001, 50; 2011, 40). In addition, the men were allowed to shop in St. Augustine and Jacksonville, Florida, and some of the prisoners may have purchased drawing materials from local mer-chants (Petersen 1971, 65, 67). The drawings by Koba and White Horse discussed in this chapter were drawn on the pages of commercial drawing books, while Bird Chief placed his drawings on the ruled pages of a book with marbled covers.

Drawings by an unidentified Kiowa prisoner, whose depictions of a courting scene and victory dance appear in this chapter, occur in a book that differs from the sketchbooks described above. The book contains ruled pages and the word "EXERCISES" is printed on its cover. The prisoners attended classes during their incarceration, and it is possible that this book was intended to be used in conjunction with one of these classes. If so, the artist may have "repurposed" the book, much as the prisoners altered the trousers issued to them, fashioning them into leggings (Lookingbill 2006, 70).

As objects of material culture, each sketchbook has its own unique history, its own unique biography. As anthropologists Arjun Appadurai (1986) and Igor Kopytoff (1986) point out, commoditization is but one possible phase in the life of an object. Many of the drawings produced at Fort Marion were purchased by tourists and reformers who visited the fort (Petersen 1971, 65; Szabo 2007, 166; 2001, 53). Lieutenant Pratt, an ardent proponent of assimilation, also presented sketchbooks to government officials and philanthropists whom he believed might aid him in his efforts at Fort Marion or in implementing his broader vision for Indian education (Szabo 2001, 51–52). However, this does not mean that the drawings were created specifically for a non-Indian audience. Indeed, the artists' fellow prisoners would have comprised the initial audience for the drawings (Berlo 1990, 137, 139; Jordan 2012, 21). Addressing the issue of the intended audience for Fort Marion drawings, Janet Berlo argues that the drawings should be viewed as "multivalent and multivocal works that hold a message for the native audience as well as the dominant culture" (Berlo 1990, 139).

It is possible to partially reconstruct the biographies of the Koba, White Horse, and Bird Chief sketchbooks. The Koba sketchbook is the best documented of the three. Several details contained in the drawings suggest that Koba created the images for a Kiowa audience. His inclusion of name glyphs in several drawings provides one such indication. While Koba's fellow Kiowa prisoners would have been able to derive an individual's identity from his name glyph, the glyph would have held little meaning for an individual from outside the Kiowa community (Jordan 2012, 21). Similarly, Koba depicts the yellow and black stripes on the south side of the *Tipi with Battle Pictures* in one of his courting scenes. A Kiowa viewing the drawing would have recognized the distinctive tipi design and its association with a prominent Kiowa family and understood that the young man depicted in the scene was courting a woman of considerable social standing. Again, such information would have been lost on non-Kiowa viewers. In 1876, Lieutenant Pratt presented the Koba sketchbook to Miss Josephine Russell. Russell was one of a number of women who volunteered as instructors at Fort Marion, providing the prisoners with lessons on spelling, grammar, phonetics, geography, and arithmetic, as well as religious instruction. Years after Russell received the sketchbook, her nephew sent it to Pratt, who added handwritten captions to the drawings (Lookingbill 2006, 107–11; Plains Indian Ledger Art 2016).

Like the Koba sketchbook, the White Horse drawings also appear to have been created with a Kiowa audience in mind. In several of the drawings, White Horse identifies himself by depicting distinctive elements of his dress and adornment, including his shield and face paint, and adding his name glyph. It is unlikely that anyone outside of the Kiowa community would have recognized the shield and face paint as belonging to White Horse. The sketchbook was collected by Noel Atwood, a twenty-three-year-old resident of St. Augustine, who presumably purchased it from White Horse. Shortly after acquiring the drawings, Atwood gave them to a cousin who resided in New York (Petersen 1971, 69).

Relatively little is known regarding the provenance of the book of drawings created by Bird Chief. Ellen Fairbanks loaned the book to the Fairbanks Museum in 1917. It had belonged to her father, Franklin Fairbanks. It is not entirely clear how he obtained the book. Fairbanks family oral history suggests that Franklin Fairbanks provided financial support for the Fort Marion prisoners and received the book as a token of appreciation. While plausible, the account cannot be corroborated. In 2006, Ellen Fairbank's heirs converted the loan into a gift to the museum (Plains Indian Ledger Art 2016).

Women and Warfare in Kiowa Society

In many Plains Indian societies, women celebrated their kinsmen's martial achievements. For example, women participated in scalp dances that marked the return of successful war parties. During the dances, women displayed war trophies, including enemy scalps, suspended from poles (Ewers 1994, 329; Grinnell 1995, 18; Stands in

Timber and Liberty 1967, 70). The Kiowa also observed this practice (Mishkin 1940, 30–31; Mooney 1979, 291–92). Yellow Wolf and Frizzle Head, veteran Kiowa warriors, emphasized that the scalp dance was held only when a party had succeeded in killing an enemy without sustaining any loss of life itself (LaBarre 1935a, 606–07, 609). An unidentified Kiowa artist, one of Bird Chief, White Horse, and Koba's fellow prisoners at Fort Marion, depicted such a celebration (Fig. 3.4). The women in the drawing wear men's headdresses and brandish lances. It was customary for Kiowa women to don their kinsman's war regalia and to carry their weapons during scalp and victory dances (Boyd 1981, 62; LaBarre 1935a, 607). Kiowa women also played important roles in warfare that extended beyond the celebration of male exploits.

Scholars have recognized nineteenth-century Plains Indian women's participation in defensive warfare, documenting instances in which women fought when their villages came under attack (Agonito and Agonito 1981, 13). Citing the almost incessant warfare that characterized Plains Indian life during much of the nineteenth

FIGURE 3.4.
Kiowa women scalp dancing to celebrate a victory. Some of the women wear their male kinsmen's headdresses and carry their lances. Unidentified Kiowa artist, ca. 1875–1878, MS 98-54. National Anthropological Archives, Smithsonian Institution, Washington, DC.

century, Medicine underscores that there was a "need for women to be assertive and able to fight for reasons of self-defense" (Medicine 1983, 275). A number of tribes sanctioned the killing of enemy women in intertribal warfare (Ewers 1994, 325–26). For example, when Cheyenne and Arapaho warriors attacked the Kiowa encampment on Wolf Creek in 1838, they killed more than fifteen Kiowa women.[3] A dozen of these women were discovered gathering plants on the outskirts of the village and were among the first individuals to be killed in the engagement (Grinnell 1995, 53–58; Hyde 1968, 79–81). Plains Indian women fared no better in their tribes' conflicts with the US military, as evidenced by female Cheyenne casualties in the Sand Creek Massacre, the Battle of the Washita, and the engagement at Middle Sappa Creek (Chalfant 1997, 136, 157–58; Grinnell 1995, 167–68, 289, 291; Hoig 1961, 152; Hyde 1968, 152, 154–55, 162–63; Stands in Timber and Liberty 1967, 169). Given this context, it is not surprising that women occasionally took up arms to defend their lives, their families, and their homes. However, rather than focusing on women's involvement in defensive fighting, Kiowa sources, including Kiowa drawings from Fort Marion, highlight women's active participation in raiding.

The Kiowa made a distinction between expeditions undertaken for the purpose of stealing enemy horses and revenge raids in which participants sought to avenge the death of a tribal member by killing an enemy. The two differed markedly, not only in regard to their objectives but also in terms of the number of personnel involved and their duration (Mishkin 1940, 28). According to Bernard Mishkin, parties rarely deviated from their stated purpose: "On the whole, there was as much singleness of objective in Kiowa war parties, as anywhere in the Plains. A revenge party seldom could be turned into a horse stealing enterprise, although capturing an enemy's horse in an attack was certainly not dishonorable. Likewise, a horse raiding expedition would prefer not to engage in combat merely for the sake of combat and scalps. However, fighting was sometimes necessary if the raiders were discovered, or if they were pursued by the victims and compelled to fight their way out" (Mishkin 1940, 31). Female participation in both types of raids is depicted in Kiowa drawings from Fort Marion. The published literature on the Kiowa documents eight specific instances of women accompanying raiding parties (Ewers 1978, 16; 1994, 328; Mishkin 1940, 48–49; Mooney 1979, 311–12;

Nye 1962, 95; 1969, 129, 153, 236–37; Rhoades 2000, 74). The field notes of the 1935 Laboratory of Anthropology field school contain four additional accounts (LaBarre 1935a, 611, 620; 1935b, 128–29). Five of the expeditions the women accompanied are identified as revenge raids (Ewers 1994, 328; LaBarre 1935a, 611, 620; Nye 1969, 129, 153, 236–37; Rhoades 2000, 74), and two are identified as horse-stealing expeditions (Mooney 1979, 311–12; Nye 1962, 95). In the remaining four cases, the nature of the raid is not specified.

While the limited number of documented cases in which Kiowa women accompanied raiding parties might be construed as evidence that this was a somewhat unusual practice, this would be a mistake. In his *Calendar History of the Kiowa Indians*, James Mooney identifies only one instance in which women accompanied a raiding party. However, his statement regarding the expedition, which took place in the spring of 1863, is revealing. He notes that, "In the early spring a large war party, accompanied by women, *as was sometimes the custom among the Kiowa*, started for Texas" (Mooney 1979, 311–12, emphasis added). Clearly, Mooney understood that this was not an isolated incident but rather an established and accepted practice in Kiowa society.[4]

The total number of women accompanying any given party was usually quite small. According to Yellow Wolf, the wife of the toyopki or leader of the expedition might accompany the party and perhaps one additional woman (LaBarre 1935b, 23). On rare occasions, the Kiowa recruited larger parties, which might consist of one hundred or more men. Women accompanied these expeditions in greater numbers. Frizzle Head recalled two such undertakings. The first was organized by Heap of Bears and was accompanied by thirteen women. The expedition culminated in a battle with the Ute in which Heap of Bears was killed. Subsequently, a large revenge raid was organized to avenge Heap of Bears and six women accompanied the raid, including a sister of the slain warrior (LaBarre 1935a, 611, 620). Kiowa drawings from Fort Marion provide additional evidence of Kiowa women's participation in horse-stealing expeditions and revenge raids.

Kiowa Women and Raiding: The Pictorial Evidence

The images Bird Chief, White Horse, and Koba created are the only extant drawings depicting Southern Plains Indian women's participation in raiding and warfare. Furthermore, Bird Chief and Koba's drawings are the only known illustrations of Plains Indian women actively engaged in the act of stealing horses. As Ewers notes, "Graphic portrayals of women warriors in action against their enemies are exceedingly rare" (Ewers 1994, 328). In his essay, Ewers identifies only two such drawings. The first, by an unidentified Cheyenne artist, depicts a woman, nude to the waist and wearing a breechclout, brandishing a carbine. The second drawing, by the Lakota artist Amos Bad Heart Bull, depicts six men and two women departing on a war party (Ewers 1994, 325–26, 328). Given that warfare was largely supplanted by other subject matter at Fort Marion, it is somewhat ironic that the majority of Plains Indian drawings depicting women's participation in warfare were produced by Kiowa artists at Fort Marion.

Wives and Husbands

Both Bird Chief and White Horse created drawings depicting couples on raiding expeditions. Given what is known regarding Kiowa women's participation in such expeditions, it is almost certain that these figures represent married couples. Almost all of the Kiowa women who travelled with raiding parties were accompanying their husbands, the sole exception being women who joined revenge parties to avenge the death of one of their close relatives. Such women acted on their own initiative (LaBarre 1935a, 589). Kiowa accounts of women participating in war expeditions specifically mention married couples. Black Bear accompanied her husband Big Bow on numerous raids (Ewers 1978, 16; LaBarre 1935a, 132). Iseeo took his wife on at least one extended raid into Chihuahua in 1853 (Nye 1962, 95). Atah joined her husband, Set-maunte, on a revenge raid in 1878 (Nye 1969, 236–37; Rhoades 2000, 74). In addition, Frizzle Head indicated that the thirteen women who accompanied the expedition led by Heap of Bears were traveling with their husbands. According to Frizzle Head, it was considered a mark of honor for wives to accompany their husbands to war (LaBarre 1935a, 589).

   A drawing by Bird Chief (Fig. 3.5) features a man and woman, likely a husband and wife, on a raiding party. The couple is depicted from the rear, while the man's horse is depicted in profile. The man is armed with a carbine or rifle, its stock visible over his right shoulder, as well as a pistol in a military-style holster. The woman wears his bow case and quiver and carries his lance. The warrior has prepared

FIGURE 3.5.
A couple stands next to a horse that has been prepared for battle. The woman wears the man's bow case and quiver and appears to be holding his lance. The man wears a pistol in a holster, and the butt of a carbine is visible over his right shoulder. Bird Chief (Kiowa), ca. 1875–1878, Bad Eye Sketchbook, Page 40. Fairbanks Museum and Planetarium, St. Johnsbury, Vermont. View the complete book at plainsledgerart.org.

his horse for battle by painting it with designs associated with his personal medicine or spiritual power and binding up its tail in red cloth (Petersen 1971, 55, 293). The treatment of the horse, as well as the armament depicted, signals the drawing's association with warfare.

During his incarceration at Fort Marion, White Horse produced a drawing (Fig. 3.6) of one of his wives accompanying him on a war expedition. In the drawing, White Horse is armed with a carbine as well as a pistol. His wife rides behind him, carrying his mountain lion skin bow case and quiver, shield, and lance. She leads his war horse, ensuring that the animal will be rested and fresh when they encounter the enemy (Petersen 1971, 80–81). Like the woman in Bird Chief's drawing, White Horse's wife is also well dressed. She wears a *màun-kàugúlhòldà* (red sleeve dress) and decorated moccasin leggings, as well as a number of bracelets. Prior to engaging the enemy, Kiowa men dressed in their finest clothing. In this manner, they prepared themselves for possible death in battle (LaBarre 1935a, 510; Petersen

FIGURE 3.6.
White Horse and his wife on a war expedition. Note that she wears his bow case and quiver and carries his lance and shield. White Horse (Kiowa), ca. 1875–1878, Joslyn Art Museum, Omaha, Nebraska.

1971, 55). This drawing suggests that Kiowa women also prepared themselves in this manner.

Revenge Raids

Koba created a two-page drawing that depicts the return of a success-ful war party (Fig. 3.7). The presence of enemy scalps and the ab-sence of captured horses indicate that the expedition was launched as a revenge raid. The caption reads "Young men returning from a foray greeted by the girls." However, a close inspection reveals that one of the members of the returning party is a woman. The figure on the far right wears a trade cloth dress and distinctive women's hide leggings. The dress is of a style known as a *màunkàugúlhòldà* and consists of a dark blue or black body with red sleeves and inset side panels (Meadows 1999, 141). Her moccasin leggings, a form of foot-wear worn exclusively by women that combined the moccasin and legging into a single article of clothing, are embellished with beadwork and conchos (Jennings 1995, 46–47; LaBarre 1935a, 507; Merrill et al. 1997, 53, 105, 109–10). The woman carries a shield indicating that she has accompanied her husband on the raid. Based on the fact that

FIGURE 3.7.
A triumphant Kiowa expedition returns to camp with two enemy scalps. Note the woman with the returning party. She wears a red sleeve dress and women's moccasin leggings. Koba (Kiowa), ca. 1875–1878, Koba-Russell Sketchbook, plate 6. Mandeville Library and Plains Indian Ledger Art Publishing Project, University of California, San Diego, La Jolla, California. View the complete book at plainsledgerart.org.

he is the only warrior depicted without a shield, the man riding next to her is likely her spouse.

The drawing depicts the party's triumphal return. The group charges the Kiowa camp, six of the men firing their revolvers in the air. This is the customary way in which victorious Kiowa parties signaled their return (LaBarre 1935a, 606–607). The party has succeeded in obtaining two enemy scalps without sustaining any casualties.[5] Two warriors riding in advance of the others display the scalps on the ends of slender poles. The trophies have been painted and decorated with eagle feathers. The members of a party who had secured an enemy scalp would paint their faces black upon their return (LaBarre 1935a, 606, 895). In Koba's drawing, the raiders' faces are painted blue. However, blue and black were often treated as interchangeable colors by Plains Indian artists (McCoy 1987, 63, 65).[6] Significantly, the woman's face is also painted. This comports with the testimony of Lone Bear, who reported that all of the members of a successful party would paint their faces, including any women in the party (LaBarre 1935b, 11).

Horse-Stealing Expeditions

Kiowa drawings produced at Fort Marion offer evidence that women also actively participated in raids on enemy horse herds. Depictions of men stealing horses frequently appear in nineteenth-century Plains Indian graphic art (Szabo 1984, 368). However, the Kiowa drawings are the only known representations of Plains Indian women engaged in stealing horses. Koba produced a two-page drawing that features a woman and two men driving a herd of stolen horses (Fig. 3.8). Koba has rendered the well-dressed woman's clothing in detail. She wears a màunkàugúlhòldà and hide moccasin leggings decorated with yellow and red ochre, lanes of beadwork, and a row of conchos. A streak of red paint applied to her face is likely associated with protective medicine. She carries both her husband's lance and shield. In all likelihood, the man who is depicted without either of these accoutrements is her husband.[7] He has prepared himself for an encounter with the enemy by painting his face and upper body yellow. The man appears well armed. He carries a carbine and wears his bow case and quiver.

The raid appears to have been extremely successful. Kiowa men who participated in horse raids later recalled that expeditions typically yielded only one or two horses per person (LaBarre 1935a, 421). In Koba's drawing, the three protagonists drive a herd of twenty-one horses and mules. While the woman shares in the party's success, she also shares in its dangers. There was always a possibility that the enemy might pursue the party and attempt to recover their stolen stock. Perhaps it is this risk that prompts the man to allow his wife

FIGURE 3.8.
A woman and two men drive a herd of stolen horses. The woman carries a man's lance and shield. Koba (Kiowa), ca. 1875–1878, Koba-Russell Sketchbook, plate 20. Mandeville Library and Plains Indian Ledger Art Publishing Project, University of California, San Diego, La Jolla, California. View the complete book at plainsledgerart.org.

to carry his shield and lance, but to keep his long range weapons near at hand.

Bird Chief created two drawings depicting Kiowa women stealing enemy horses. In one (Fig. 3.9), a mounted couple drives a stolen horse ahead of them. The woman is shown in the lead. She wears her husband's bow case and quiver and carries his shield and lance, along with a rifle. Her husband, armed with a second carbine or rifle, brings up the rear. A close examination of his firearm reveals that the hammer on the percussion lock is cocked, meaning that the weapon is ready to fire. As in the drawing by Koba, the weapon held at the ready conveys the inherent risks involved in such an enterprise.

In the second drawing (Fig. 3.10), a solitary woman herds a stolen horse.[8] Two pieces of evidence indicate that this is a scene from a horse raid. First, the woman carries her husband's shield and bow case and is armed with a firearm. Such behavior only makes sense within the context of a raid. Horse tack provides another important clue. Elsewhere in the sketchbook, Bird Chief depicts women leading

FIGURE 3.9.
A woman and man, likely a wife and husband, herd a stolen horse. The man is armed with a carbine or rifle, as is the woman. In addition, she wears the man's bow case and quiver and carries his shield and lance. Bird Chief (Kiowa), ca. 1875–1878, Bad Eye Sketchbook, Page 25. Fairbanks Museum and Planetarium, St. Johnsbury, Vermont. View the complete book at plainsledgerart.org.

FIGURE 3.10.
A woman herds a stolen horse. She carries a carbine, shield, and lance. Bird Chief (Kiowa), ca. 1875–1878, Bad Eye Sketchbook, Page 23. Fairbanks Museum and Planetarium, St. Johnsbury, Vermont. View the complete book at plainsledgerart.org.

pack horses and extra mounts. The distinctive halter and the conspicuous absence of a lead rope in this drawing indicate that this is a captured horse that is being herded rather than led along. Bird Chief appears to have employed this style of halter as a convention to denote enemy horses. An analysis of his drawings reveals that he depicts all of the Kiowa individuals' mounts as well as all but one the horses that they lead, wearing silver-decorated bridles.

Combat

Kiowa women who accompanied raiding parties typically did not engage in combat (LaBarre 1935a, 589). Nevertheless, exceptional circumstances sometimes dictated that women take up arms. In one celebrated incident, Zepkoyette (Big Bow) and his wife Black Bear were surrounded by a detachment of Mexican soldiers. Having held the troops at bay all day, the couple broke through the enemy lines during the night and made their escape. Black Bear, armed with a revolver, fought alongside her husband, who was wounded several times during the fighting. For many years, the deed was commemorated on the *Tipi with Battle Pictures*, the Kiowa tipi that featured depictions of martial exploits (Ewers 1978, 16; LaBarre 1935a, 132). Keintaddle, a Kiowa woman who recounted Black Bear's story for ethnographer Jane Richardson, indicated that this was not an isolated incident. According to Keintaddle, Black Bear had taken an active role in the fighting during other expeditions as well (LaBarre 1935a, 132).

While there are no known Kiowa drawings depicting women engaged in combat, a drawing by Bird Chief depicts a woman who appears ready to fight should the need arise (Fig. 3.11). It is the last drawing in a sequence of three illustrating a horse-stealing episode. The first drawing depicts a dismounted man who has climbed a hill below which two horses and a mule are picketed. In the next drawing, the man, now mounted, drives two horses ahead of him. The final drawing in the sequence features a woman sitting atop a knoll armed with a rifle or carbine. A dashed line leading from a horse to her

FIGURE 3.11.
A woman stands watch atop a hill. She is armed with a rifle or carbine. The presence of a man's shield and weapons suggests that she is awaiting her husband's return. Bird Chief, (Kiowa), ca. 1875–1878, Bad Eye Sketchbook, Page 37. Fairbanks Museum and Planetarium, St. Johnsbury, Vermont. View the complete book at plainsledgerart.org.

position shows that she has dismounted, secured her horse, and climbed the eminence, likely seeking a view of the surrounding area. A man's quiver and bow case, as well as his shield, are suspended from a tree, and his lance stands close by. The presence of the shield and weapons indicates that the woman is awaiting her husband's return, while the fact that the woman is armed and has sought out a commanding vantage point suggests potential danger. When the image is viewed as part of a sequence of drawings rather than independently, it becomes clear that it represents a woman waiting for her husband to return with stolen horses. Atop the rise she is in a position to see if he is being pursued by the enemy. Realizing that this is a possibility, she has armed herself.

Understanding Women's Roles in Warfare

Pictorial and archival sources indicate that Kiowa women routinely accompanied horse raiding and revenge parties. They also provide a basis for reconsidering the nature of women's participation in these expeditions. Wilbur S. Nye relates Andrew Stumbling Bear's account of a Kiowa raid into Chihuahua in the fall of 1853 in which Iseeo and his wife participated. Describing women's roles on such expeditions, Nye states, "It was their duty to care for the extra horses, cook the food, and remain in charge of the raid headquarters while the warriors were on the prowl" (Nye 1962, 94–95). Nye's depiction of Kiowa women reflects the widespread representation of Plains Indian women as drudges who performed menial chores while their husbands engaged in hunting and warfare (Medicine 1983, 267; Weist 1983). Whether Nye's description of women's responsibilities while on raids is based on testimony from Kiowa consultants or his own assumptions regarding Kiowa gender roles is unclear. However, there is cause to question Nye's characterization of the role Kiowa women played on such expeditions.

According to Frizzle Head, Kiowa women who accompanied raiding parties were not expected to work. While on a raid, a woman enjoyed a respite from her usual tasks, and men assumed responsibility for food preparation and the care of the horses (LaBarre 1935a, 589–90). Mary Buffalo concurred, explaining to LaBarre that while on a raid the men would "treat [the] woman nice, she no work, every man nice to her, she not work, man work" (LaBarre 1935b, 23). Weasel Tail, a Blood warrior, identified an almost identical arrangement among his own people. He reported that his wife, who accompanied him on five raids, never performed camp chores, as these were assigned to teenage boys (Ewers 1994, 330). Since this pattern is documented on both the Northern and Southern Plains, it is possible that it existed not only among the Kiowa and Blood but among other tribes as well.

Regardless of whether Kiowa women who accompanied raiding parties occasionally cooked meals or cared for the expeditions' horses, there is sufficient evidence to dispute Nye's suggestion that these women confined themselves to the raiding parties' temporary camps. Sources indicate that some Kiowa women were not satisfied to remain in the relative safety of these camps and chose instead to be present when their parties encountered the enemy. Two individuals interviewed by members of the 1935 field school reported that women who accompanied raiding parties would encourage the men during engagements by ululating. Indeed, Keintaddle recalled the warrior Big Bow stating that hearing his wife ululate in combat strengthened his courage and boosted his confidence (LaBarre 1935b, 23; 1935a, 132).[9] In order to be heard by the men, these women must have been extremely close to, if not in the midst of, the fighting.

Further evidence for women's presence in combat is offered by Frizzle Head, who recounted a revenge raid against the Utes. The raid was undertaken to avenge Heap of Bear's death at the hands of Ute warriors during the preceding year. According to Frizzle Head, six women participated in the revenge raid, including one of Heap of Bear's sisters. On the return leg of the expedition, the party attacked a group of Utes and one of the Kiowa women was wounded during the ensuing engagement (LaBarre 1935a, 611).

Similarly, Atah accompanied her husband on a revenge raid in 1878 that culminated in the killing of an Anglo man near Quanah, Texas. She later recalled that she had not been present when the man was killed, explaining that she was mounted on a slow horse and could not keep up with the men in the party (Nye 1962, 231–32; 1969, 236–37; Rhoades 2000, 74). Her statement is significant because it reveals that she was not tending camp that day but rather riding alongside the rest of the party, as they traversed the countryside searching for the enemy.

Motivational Factors

The 1935 field notes also make a major contribution to our understanding of the motivations that prompted Plains Indian women's
participation in warfare. In her influential essay on Plains Indian
women and warfare, Medicine laments that, "The motivation for
women to engage in war can only be conjectured" (Medicine 1983,
274). However, Kiowa consultants clearly delineated Kiowa women's
motivations for joining raiding parties. The significance of their testimony is twofold. First, it provides evidence for evaluating earlier
scholarly speculation regarding women's motivations for engaging
in warfare. More importantly, it elucidates the connection between
a specific form of marriage and Kiowa women's participation in war
expeditions, revealing a previously unrecognized motivation for
women to join raiding parties.

Elopement and Raiding

Kiowa consultants observed that when a woman decided to desert her
husband for another man, the new couple might join an expedition
bound for enemy territory. According to Mrs. Big Bow, a woman who
wished to leave her husband would bide her time, waiting until there
was a raiding party about to depart, and then press her paramour to
join the party and take her with him (LaBarre 1935a, 345). Indeed,
Frizzle Head identified absconding with a lover as one of the principle reasons that a woman might choose to accompany an expedition
(LaBarre 1935a, 589). Weston LaBarre referred to this practice as "war
elopement" (LaBarre 1935a, 181). The 1935 field notes document at
least three specific cases that conform to this pattern. Interestingly,
one of the incidents involved Bird Chief, who created several of the
drawings discussed in this chapter (LaBarre 1935a, 198, 216, 321).

Bird Chief gained fame by absconding with one of Heap of Bear's
wives not once, but twice. The first time, the couple immediately
departed on a raiding party. Shortly after their return, Heap of
Bears retaliated against them. He attempted to kill Bird Chief but
succeeded only in severely wounding him. He also reclaimed his wife,
punishing her for her infidelity by gashing her nose with a knife.
After he had recovered, Bird Chief fled with the woman a second
time. Again, they joined a raiding party. However, this time, the

couple stayed gone for three years, returning only after Heap of Bear's death (LaBarre 1935a, 216; Mishkin 1940, 48–49).[10]

Enticing another man's wife to leave him was considered an act of bravado. Lower ranking Kiowa men frequently sought to elevate their status by shaming or insulting higher ranking men, a practice that Mishkin refers to as "outfacing." As in the case above, outfacing often took the form of taking a higher ranking man's wife. By engaging in such a deliberate provocation, a man sought to demonstrate his fearlessness and courage. Success in such an endeavor earned a man accolades and elevated his social status. With regard to the incident involving Bird Chief, it is important to note that he had been born into the lowly *kaan* class. Prior to the incident involving Heap of Bears, Bird Chief had amassed both martial honors and material wealth, thereby elevating his status. Nevertheless, some elements of Kiowa society refused to recognize him as a member of the prestigious *onde* rank, citing his modest origins. Bird Chief's decision to steal the wife of Heap of Bears, a prominent member of the onde rank and a residence band leader, represented an attempt to elevate and solidify his own standing in Kiowa society. By his actions, Bird Chief demonstrated that he was unafraid of Heap of Bears (Mishkin 1940, 48–49).

Precisely because it constituted an attack on her husband's honor, stealing a married woman was an enterprise fraught with risk and both the man and woman involved had cause to anticipate retribution from the affronted husband (Greene 2001, 210). By joining a raiding party, a woman and her paramour attempted to escape her husband's wrath. A man whose wife abandoned him for another was expected to take swift and decisive action. In nineteenth-century Kiowa society, the use of violence to punish a wayward wife and compel her to return to her marriage was not only condoned but expected (LaBarre 1935a, 180, 200–202, 216). Mary Buffalo, a Kiowa elder explained that, "If a man's wife wrongs him, he is expected to do something drastic about it or [else] lose face" (LaBarre 1935a, 180).[11] Women who left their husbands almost always faced retaliation. Aggrieved husbands physically assaulted their estranged wives or attempted to disfigure them by cutting off their noses (LaBarre 1935a, 162, 240, 342; Mooney 1979, 233, 353; Parsons 1929, 136). In rare cases, distraught husbands killed their former spouses (Mooney 1979, 340–41; Parsons 1929, 135).[12]

As the case of Bird Chief makes clear, a man who had stolen a married woman had reason to be concerned for his own safety. A confrontation with the woman's husband was almost unavoidable. A husband was expected to challenge his wife's lover and failure to do so elicited public ridicule and accusations of cowardice (LaBarre 1935a, 200, 216). Slighted husbands typically targeted their rival's property, either seizing or destroying it, and retribution frequently took the form of killing or appropriating a number of the offender's horses (LaBarre 1935a, 5, 98, 342; Mooney 1979, 337, 353; Parsons 1929, 136).[13] However, there was also a risk that the husband would attempt to recoup his honor by beating or even killing his rival (LaBarre 1935a, 162, 200; 342). By joining a raiding party and absenting themselves from camp, the couple attempted to avoid a violent confrontation with the woman's husband. In theory, the duration of the expedition functioned as a cooling-off period during which the affronted husband's desire for revenge might abate.

Revenge

Another motivation that prompted Kiowa women to join war expeditions was the desire to avenge a relative who had died at the hands of the enemy. When he was interviewed by Mishkin in 1935, Frizzle Head, an elderly veteran of intertribal warfare and clashes with the US military, stated that women sometimes joined revenge parties in order to avenge a deceased relative (LaBarre 1935a, 589–90). Following Heap of Bear's death at the hands of the Ute, the Kiowa organized a large revenge raid. Among the six women who joined this expedition was Heap of Bear's sister. After the party had killed a Ute, she requested the slain man's scalp. Each night on the return journey, she and the other women held a scalp dance to celebrate their victory (LaBarre 1935a, 611).[14]

Oral narratives collected from members of the Crow and Blackfoot tribes indicate that the desire for revenge also motivated some Northern Plains women. Pretty Shield, a Crow woman who worked closely with Frank Bird Linderman, recalled the battlefield exploits of a young Crow woman named Other Magpie. Eager to avenge the death of her brother, who had been killed by the Lakota, Other Magpie joined the Crow auxiliaries recruited by General Crook and distinguished herself during the Battle of the Rosebud on June 17, 1876.

When Lakota and Cheyenne warriors attacked Crook's camp, she entered the fray armed only with a coup stick and a knife. During the engagement, Other Magpie counted coup on a Lakota warrior, whom she later scalped (Ewers 1965, 10–11; Linderman 1972, 228–30).

Running Eagle, a Blackfoot woman, embarked on her career as a warrior after her husband was killed in an engagement with the Crow. Desirous of revenge, she prayed to the sun for medicine or spiritual power that would grant her success in war. According to Running Eagle, she received the power that she sought. She led a number of successful raids but was ultimately killed while attempting to steal picketed horses from a Flathead village around 1860 (Ewers 1965, 12–13). Combined with the accounts of Running Eagle and Other Magpie, Frizzle Head's testimony regarding Kiowa women's participation in revenge raids suggests that the pattern of women going to war to avenge their deceased kin existed throughout the Plains.

Honor

Medicine argues that some Plains Indian women who engaged in warfare were likely motivated by the desire to gain prestige by distinguishing themselves through martial exploits, just as male warriors sought to accrue war honors and thereby bolster their social standing (Medicine 1983, 274–75). This was certainly true among the Kiowa. In nineteenth-century Kiowa society, women's participation in war expeditions was both recognized and honored. Women who accompanied their husbands on war expeditions or joined revenge parties earned their community's respect (LaBarre 1935a, 589).

The Kiowa recognized a variety of war-related acts to which they attached varying levels of prestige. In 1935, Kuito provided a list of deeds that ranked as war honors. Included on his list are horse stealing, regular participation in war expeditions, and conducting oneself properly while on a raid, both in camp and on the march (LaBarre 1935a, 628). While these acts paled in comparison to rescuing a comrade in battle or counting first coup on an enemy, their performance was nevertheless considered admirable and earned men a measure of respect (Mishkin 1940, 39–40). Kiowa women who regularly accompanied their husbands on raids, aided in the capture of enemy horses, or functioned as effective members of raiding parties likewise garnered prestige. Among the Cheyenne, women who had accompanied their husbands on war expeditions were afforded special status. These women formed an exclusive sorority that held meetings similar to those conducted by male warrior societies (Powell 1981, 134). While there is no evidence that Kiowa women who accompanied their husbands to war formed similar societies, women did avail themselves of other culturally sanctioned means of calling attention to their participation in martial endeavors.

Kiowa men commemorated their brave deeds by bestowing names that referenced their martial exploits on their children or grandchildren (Greene 1996b, 222; LaBarre 1935a, 627; Nye 1969, 199–200). Through her participation in the 1878 revenge raid, Atah appears to have earned the same prerogative. Atah bestowed two names on her granddaughter Evelyn Longhorn that were inspired by her involvement in the raid. These names were Tsoimahkoongyah (Coffee Grinder), a reference to the war trophy that she claimed from a slain Texan's wagon, and Gyahhonedomei (Went on the Last Raid) (Rhoades 2000, 74). Atah's experience indicates that Kiowa women, like Kiowa men, employed naming as a means of honoring and commemorating their participation in warfare.

Finally, the drawings themselves evidence Kiowa men's willingness to mark and commemorate female participation in raiding and warfare. This stands in contrast to Crow society, in which men appear to have been reluctant to celebrate female participation in warfare. For example, Linderman notes that none of the Crow men whom he interviewed recounted the story of Other Magpie, the Crow woman who fought at the Battle of the Rosebud. It was not until he interviewed Pretty Shield, a Crow woman, that he learned of Other Magpie's exploits (Linderman 1972, 227–28). Indeed, Pretty Shield's testimony suggests that Crow men were reluctant to discuss Other Magpie; she explained to Linderman that, "[T]hey do not like to tell of it" (Linderman 1972, 228). In contrast, Kiowa men sought to depict and document women's contributions to warfare in their drawings. White Horse, Bird Chief, and Koba could have easily omitted women from their depictions of raids and horse-stealing enterprises. However, they chose not to do so, opting instead to include them and to render them in as much detail as the male protagonists.

Conclusion

Analysis of Kiowa women's participation in raiding and warfare, as documented in the drawings produced at Fort Marion and in the 1935 interviews, serves to expand our understanding of Plains Indian women's involvement in warfare. A comparison of the Kiowa data with accounts drawn from other Plains Indian societies reveals significant variation in the roles women played in warfare. Medicine notes that many historic accounts of female warriors "do not distinguish between women who pursued warfare as a life occupation and those who joined war parties on a situational basis" (Medicine 1983, 274). The Kiowa data help bring this distinction into sharper focus.

Citing ethnohistorical accounts drawn from Northern Plains tribes, Medicine (1983, 273–74) identifies women who eschewed traditional female gender roles and chose instead to devote their lives to distinguishing themselves in the arena of warfare. Medicine characterizes these individuals as "women who pursued warfare as an extension of their manly inclinations" (Medicine 1983, 274). These individuals adopted the role of the "warrior woman" and sought distinction "outside their customary sex role assignments" (Medicine 1983, 269). Becoming a warrior woman entailed rejecting ascribed female gender roles and embracing an alternative model.

At least one Piegan woman, Running Eagle, pursued an alternative gender role that combined masculine and feminine elements. Running Eagle gained notoriety for leading raids, including horse-stealing expeditions against the Flathead. Yet while on these expeditions she also performed female tasks such as cooking and mending clothing. Her manner of dress on raids reflected her hybrid identity, combining elements of male and female clothing. Running Eagle, like the women Medicine discusses, appears to have pursued warfare as a vocation (Ewers 1994, 328).

The Kiowa case provides a different perspective on women's participation in warfare. Drawings depicting Kiowa women's participation in raiding parties should not be construed as evidence of the existence of an established "warrior woman" role in nineteenth-century Kiowa society. None of the Kiowa accounts in the 1935 field notes treat female participation in raiding and warfare as an example of women asserting or claiming an alternative gender role, neither do they provide any indication that an alternative gender role equivalent to Medicine's (Medicine 1983, 273) "warrior woman" existed among nineteenth-century Kiowa society. Indeed, there is no evidence that the Kiowa recognized such an institution.

In Bird Chief's drawings, depictions of women accompanying raiding and horse-stealing parties appear alongside drawings of women caring for children and moving camp, activities associated with female gender roles in Kiowa society (LaBarre 1935a, 105, 350; Mayhall 1962, 115). The coexistence of martial and domestic scenes suggests that Kiowa women's participation in raiding and warfare did not represent a repudiation of established gender norms. Quite the contrary, it suggests that participation in such endeavors was one of the ascribed and socially sanctioned behaviors for Kiowa women.

Furthermore, in Kiowa narratives, women's participation in war expeditions is situated within the context of female kinship obligations. It is in their role as wives that Kiowa women most often accompanied men on raids. Such behavior on the part of a wife was considered laudable. In addition, women joined revenge parties to avenge the deaths of close kinsmen, particularly their fathers and brothers. Honoring a deceased relative by accompanying a revenge raid also earned women admiration (LaBarre 1935a, 589). Far from marking a departure from women's ascribed roles as wives, daughters, and sisters, female participation in war expeditions represented an extreme form of familial devotion.

The Kiowa practice of women accompanying their husbands on raids closely resembles a pattern documented among the Blackfoot (Medicine 1983, 274). Ewers (1994, 330) recorded several accounts of Blood and Piegan women who went with their husbands on war expeditions. Weasel Tail, a Blood elder, reported that, "A lot of old-timers took their wives on war parties. Their wives wanted to go" (Ewers 1994, 330). Indeed, he reported that his own wife had accompanied him in five separate engagements with the enemy (Ewers 1994, 330). As in the Kiowa case, there is no indication that these Blackfoot wives' participation in warfare was viewed or construed as a renunciation of established female gender roles or an attempt to adopt an alternative gender role.

Finally, the Kiowa drawings and the 1935 field notes also offer a clearer picture of the role Kiowa women played on war expeditions. Accounts of Plains Indian women leading war expeditions are extremely rare and are confined to the Northern Plains. The practice is documented among the Dakota, Crow, and Blackfoot (Ewers 1965, 12–13; 1994, 328–29; Medicine 1983, 272, 274). While Kiowa women routinely accompanied raiding parties, there is no evidence suggesting that they organized or directed these enterprises. The field notes of the 1935 Laboratory of Anthropology field school contain detailed discussions of the qualifications for serving as a *toyopki* or leader of a war party, as well as the responsibilities that accompanied the position (LaBarre 1935a, 592–94). However, they provide no indication that women served in this capacity. Given that the Kiowa consultants interviewed in 1935 volunteered accounts of Kiowa women's participation in warfare, it is unlikely that they would have omitted information regarding female leadership in this arena.

Bird Chief, Koba, and White Horse's drawings, as well as the 1935 field notes, reveal the extent of Kiowa women's participation in raiding and warfare. Women who accompanied raiding parties emerge as full partners in these undertakings. Their presence on the battlefield and the encouragement that they offered their husbands during engagements served to bolster the latter's morale. Women also actively participated in horse raids, helping to drive the stolen stock. And, although they do not appear to have regularly engaged in combat, when necessary, women contributed to the defense of their parties by arming themselves. All the while, these women shared in the risks inherent in raiding and warfare. By exposing themselves to these dangers, Kiowa women earned the approbation and respect of their communities.

Notes

1. For other examples of Cheyenne and Lakota drawings depicting men killing enemy women, see Berlo (1996, 102–103, 208–209). For an Arapaho drawing of a man shooting an enemy woman, see Petersen (1990, XXXIII).

2. The book of drawings by Bird Chief is in the collection of the Fairbanks Museum and Planetarium, St. Johnsbury, Vermont. The White Horse drawings are in the collection of the Joslyn Art Museum, Omaha, Nebraska. The Koba-Russell Sketchbook is one of a number of Plains Indian drawing books held by the Mandeville Library and Plains Indian Ledger Art Publishing Project at the University of California, San Diego.

3. Cheyenne sources acknowledge the contribution that the Kiowa and Apache women made to the defense of the village. As the attack unfolded, the women felled trees and quickly constructed a hasty barricade around the village. The barricade prevented the Cheyenne and Arapaho from making mounted

charges into the camp and provided cover from behind which the Kiowa and their allies could fight (Hyde 1968, 81).

4. Far from indicating that women's participation in raiding parties was rare, the fact that Kiowa calendar keepers did not record other instances of this practice suggests that it was a relatively common occurrence. Calendar keepers selected especially memorable events to represent each season, events that were somehow unusual or out of the ordinary (Greene 2001, 24, 163). Accordingly, one would not expect to find commonplace events recorded in the calendars. It is worth noting that it was not the presence of women within the raiding party's ranks that made the 1863 expedition noteworthy but rather an acoustical phenomenon that the party encountered. One day, as the party started singing, the tree tops seemed to echo their song, an occurrence that the members of the party attributed to spiritual forces. This unusual experience prompted one calendar keeper to designate the winter of 1862–1863 as the "Tree Top Winter" (Mooney 1979, 311–12).

5. When a member of a war party was killed, there was no triumphal entry into camp, regardless of whether the party had succeeded in securing an enemy scalp. The members of the party would simply filter back into the camp (LaBarre 1935a, 608).

6. Two Kiowa drawings of parties returning with scalps appear in the Julian Scott ledger. In both of these drawings, as in Koba's, the individuals are depicted with blue rather than black face paint (McCoy 1987, 36, 49, 63, 65).

7. An attempt to identify the shield that the woman carries proved unsuccessful. The design does not match any of the Kiowa shield models collected by James Mooney (1904), nor does the design appear in any of the sketches of Kiowa shields collected by Mooney (Smithsonian Institution 2014).

8. A comparison of this woman with the female figure in the preceding drawing reveals differences in their dress and horse tack. In addition, the shields that the women carry also differ. These inconsistencies indicate that the drawings depict different episodes and possibly depict two different women.

9. The name appears as Kintadl in the 1935 field notes. However, Keintaddle is the spelling preferred by her descendants and which will be utilized here (Jennings 2000, 95; Rhoades 2000, 89).

10. While some Kiowa raiding parties are documented as having remained in the field for up to two years, it seems likely that the couple joined a series of successive expeditions (Mishkin 1940, 28). When the members of a returning party encountered an outbound expedition, they were free to join it (LaBarre 1935a, 594).

11. According to Old Man Horse, one of John Collier's informants, Sun Boy's reputation suffered because he failed to act when his wife left him for another man. Rather than confronting or punishing the couple, Sun Boy meekly acquiesced, allowing his wife to take up residence with the man. He was roundly condemned for his inaction, which people interpreted as a sign that he was afraid of the other man (LaBarre 1935a, 200).

12. A Silver Horn drawing in the collection of the Nelson-Atkins Museum of Art depicts a Kiowa husband who, having pursued and caught his wife and her lover, has shot her in the head, killing her. The caption reads "Chane-doodle (Kiowa)

overtaken after eloping with the wife of Tau-dome, Tau-dome killed his wife and takes all of Chane-doodle's clothing and ponies" (Greene 2001, 212–13).

13. For example, when Black Buffalo stole Appearing Wolf's wife during the 1873 Sun Dance, the latter responded by shooting seven of Black Buffalo's horses and seizing several others (Mooney 1979, 337).

14. Citing an account in Wilbur S. Nye's *Bad Medicine and Good: Tales of the Kiowas*, Ewers describes a Kiowa woman accompanying a "party of thirty-seven men to avenge an enemy killing of her husband" (Ewers 1994, 328). Ewers claim that the woman seeking vengeance for her deceased husband is incorrect. A review of Nye reveals that Ewers is referring to a woman named Atah and her participation in the 1879 revenge raid carried out in retaliation for the Texas Ranger's killing of White Cow Bird (Nye 1962, 227–32). However, Atah was not married to White Cow Bird, but rather to Set-maunte (aka Setmauntay), one of White Cow Bird's brothers. While Atah joined Set-maunte on the raid, it remains unclear whether her decision was motivated by a desire to avenge the death of her brother-in-law or if she was following the established pattern of wives accompanying their husbands on war expeditions (Nye 1962, 228; 1969, 237; Rhoades 2000, 74).

References

Agonito, Rosemarry, and Joseph Agonito. 1981. "Resurrecting History's Forgotten Women: A Case Study from the Cheyenne Indians." *Frontiers: A Journal of Women Studies* 6: 8–16.

Appadurai, Arjun. 1986. "Introduction: Commodities and the Politics of Value." In *The Social Life of Things: Commodities in Cultural Perspective*, edited by Arjun Appadurai, 3–63. Cambridge: Cambridge University Press.

Berlo, Janet C. 1982. "Wo-Haw's Notebooks: 19th Century Kiowa Indian Drawings in the Collections of the Missouri Historical Society." *Gateway Heritage* 3(2): 2–13.

———. 1990. "Portraits of Dispossession in Plains Indian and Inuit Graphic Arts." *Art Journal* 49: 133–41.

———, ed. 1996. *Plains Indian Drawings 1865–1935: Pages from a Visual History.* New York: Harry N. Abrams.

Boyd, Maurice. 1981. *Kiowa Voices, Vol. 1: Ceremonial Dance, Ritual, and Song.* Fort Worth: Texas Christian University Press.

Calloway, Colin G., ed. 2012. *Ledger Narratives: The Plains Indian Drawings in the Mark Lansburgh Collection at Dartmouth College.* Norman: University of Oklahoma Press.

Chalfant, William Y. 1997. *Cheyennes at Dark Water Creek: The Last Fight of the Red River War.* Norman: University of Oklahoma Press.

Ewers, John C. 1965. "Deadlier than the Male." *American Heritage* 16: 10–13.

———. 1978. *Murals in the Round: Painted Tipis of the Kiowa and Kiowa-Apache Indians.* Washington, DC: Smithsonian Institution Press.

———. 1994. "Women's Roles in Plains Indian Warfare." In *Skeletal Biology in the Great Plains*, edited by Douglas W. Owsley and Richard L. Jantz, 325–32. Washington, DC: Smithsonian Institution Press.

Greene, Candace S. 1985. "Women, Bison, and Coups: A Structural Analysis of Cheyenne Pictographic Art." PhD diss., University of Oklahoma.

———. 1993. "The Tepee with Battle Pictures." *Natural History* 102(10): 68–76.

———. 1996a. "Structure and Meaning in Cheyenne Ledger Art." In *Plains Indian Drawings 1865–1935: Pages from a Visual History,* edited by Janet Catherine Berlo, 26–33. New York: Harry N. Abrams.

———. 1996b. "Exploring the Three 'Little Bluffs' of the Kiowa." *Plains Anthropologist* 41: 221–42.

———. 1997. "Southern Plains Graphic Art before the Reservation." *American Indian Art Magazine* 22(3): 44–53.

———. 2001. *Silver Horn: Master Illustrator of the Kiowas.* Norman: University of Oklahoma Press.

———. 2004. "From Bison Robes to Ledgers: Changing Contexts in Plains Drawings." *European Review of Native American Studies* 18: 21–29.

Greene, Candace S., and Thomas D. Drescher. 1994. "The Tipi with Battle Pictures: The Kiowa Tradition of Intangible Property Rights." *Trademark Reporter* 84: 418–33.

Grinnell, George B. 1897. *Manuscript 1987-d Data on Cheyenne Painted Shields and Lodges, National Anthropological Archives,* Smithsonian Institution Press, Washington, DC.

———. 1910. "Coup and Scalp among the Plains Indians." *American Anthropologist* 12: 296–310.

———. 1995. *The Fighting Cheyenne.* North Dighton, MA: JG Press.

Harris, Moira F. 1989. *Between Two Cultures: Kiowa Art from Fort Marion.* St. Paul, MN: Pogo Press.

Hoig, Stan. 1961. *The Sand Creek Massacre.* Norman: University of Oklahoma Press.

Hyde, George E. 1968. *Life of George Bent: Written from His Letters.* Edited by Savoie Lottinville. Norman: University of Oklahoma Press.

Jennings, Vanessa Paukeigope. 1995. "Kiowa Beadwork." *Gilcrease Journal* 3(2): 44–50.

———. 2000. "Why I Make Cradles." In *Gifts of Pride and Love: Kiowa and Comanche Cradles,* edited by Barbara A. Hail, 71–76. Bristol, RI: Haffenreffer Museum of Anthropology.

Jordan, Michael Paul. 2012. "Striving for Recognition: Ledger Drawings and the Construction and Maintenance of Social Status during the Reservation Period." In *Ledger Narratives: The Plains Indian Drawings of the Lansburgh Collection at Dartmouth College,* edited by Colin G. Calloway, 20–33. Norman: University of Oklahoma Press.

Jordan, Michael Paul, and Daniel C. Swan. 2011. "Painting a New Battle Tipi: Public Art, Intellectual Property, and Heritage Construction in a Contemporary Native American Community." *Plains Anthropologist* 56: 195–213.

Keim, De Benneville Randolph. 1885. *Sheridan's Troopers on the Borders: A Winter Campaign on the Plains.* Philadelphia: David McKay.

Keyser, James D., Linea Sundstrom, and George Poetschat. 2006. "Women in War: Gender in Plains Biographic Rock Art." *Plains Anthropologist* 51: 51–70.

Kopytoff, Igor. 1986. "The Cultural Biography of Things: Commoditization as Process." In *The Social Life of Things: Commodities in Cultural Perspective,* edited by Arjun Appadurai, 64–91. Cambridge: Cambridge University Press.

LaBarre, Weston. 1935a. Typescript of student's notes (combined notes of William Bascom, Donald Collier, Weston LaBarre, Bernard Mishkin, and Jane Richardson of the 1935 Laboratory of Anthropology Field School, led by Alexander Lesser), Papers of Weston LaBarre, National Anthropological Archives, Smithsonian Institution, Washington, DC.

———. 1935b. Personal fieldnotes of Weston LaBarre from the 1935 Laboratory of Anthropology Field School, led by Alexander Lesser, Alexander Lesser Collection. American Indian Studies Research Institute, Indiana University, Bloomington.

Linderman, Frank B. 1972. *Pretty-Shield: Medicine Woman of the Crows*. Lincoln: University of Nebraska Press.

Lookingbill, Brad D. 2006. *War Dance at Fort Marion: Plains Indian War Prisoners.* Norman: University of Oklahoma Press.

Mayhall, Mildred P. 1962. *The Kiowa.* Norman: University of Oklahoma Press.

McCoy, Ronald. 1987. *Kiowa Memories: Images from Indian Territory, 1880.* Santa Fe, NM: Morning Star Gallery.

McLaughlin, Castle. 2013. *A Lakota War Book from the Little Bighorn: The Pictographic "Autobiography of Half Moon."* Cambridge, MA: Peabody Museum Press.

Meadows, William C. 1999. *Kiowa, Apache, and Comanche Military Societies: Enduring Veterans 1800 to the Present.* Austin: University of Texas Press.

Medicine, Beatrice. 1983. "Warrior Women—Sex Role Alternatives for Plains Indian Women." In *The Hidden Half: Studies of Plains Indian Women*, edited by Patricia Albers and Beatrice Medicine, 267–80. New York: University Press of America.

Mishkin, Bernard. 1940 (1992). *Rank and Warfare Among the Plains Indians.* Monographs of the American Ethnological Society, vol. 3. Seattle: University of Washington Press. Reprint, Lincoln: University of Nebraska Press.

Merrill, William L., Marian Kaulaity Hansson, Candace S. Greene, and Frederick J. Reuss. 1997. *A Guide to the Kiowa Collections at the Smithsonian Institution.* Smithsonian Contributions to Anthropology, vol. 40. Washington, DC: Smithsonian Institution Press.

Mooney, James. 1904. "James Mooney and Silver Horn Notebook of Kiowa Shield Designs." Manuscript 2531: Volume 12, National Anthropological Archives, Smithsonian Institution, Washington, DC.

———. 1979 (1898). *Calendar History of the Kiowa Indians.* Washington, DC: Bureau of American Ethnology. Reprint, Washington, DC: Smithsonian Institution Press.

Nye, Wilbur S. 1962. *Bad Medicine and Good: Tales of the Kiowas.* Norman: University of Oklahoma Press.

———. 1968. *Plains Indian Raiders: The Final Phases of Warfare from the Arkansas to the Red River.* Norman: University of Oklahoma Press.

———. 1969. *Carbine and Lance: The Story of Old Fort Sill.* Norman: University of Oklahoma Press.

Parsons, Elsie Clews. 1929. *Kiowa Tales.* New York: American Folk-Lore Society.

Petersen, Karen D. 1971. *Plains Indian Art from Fort Marion.* Norman: University of Oklahoma Press.

———. 1990. *The Edwards Ledger Drawings: Folk Art by Arapaho Warriors.* New York: David A. Schorsch.

Plains Indian Ledger Art Publishing Project. "Bad Eye Ledger—Provenance." Accessed March 1, 2016. https://plainsledgerart.org/plates/view/486.

Powell, Peter J. 1981. *People of the Sacred Mountain.* 2 vols. New York: HarperCollins College Division.

Pratt, Richard Henry. 2003. *Battlefield and Classroom: Four Decades with the American Indian, 1867-1904.* Norman: University of Oklahoma Press.

Rhoades, Bernadine Herwona Toyebo. 2000. "Memories of Atah." In *Gifts of Pride and Love: Kiowa and Comanche Cradles,* edited by Barbara A. Hail, 71–76. Bristol, RI: Haffenreffer Museum of Anthropology.

Smith, Marian W. 1938. "The War Complex of the Plains Indians." *Proceedings of the American Philosophical Society.* 78: 425–64.

Smithsonian Institution. 2014. "Collections Search Center." Accessed November 30, 2015. http://collections.si.edu/search/.

Stands In Timber, John, and Margot Liberty. 1967. *Cheyenne Memories.* Lincoln: University of Nebraska Press.

Szabo, Joyce M. 1984. "Howling Wolf: A Plains Artist in Transition." *Art Journal* 44: 367–73.

———. 1994. *Howling Wolf and the History of Ledger Art.* Albuquerque: University of New Mexico Press.

———. 2001. "From General Souvenir to Personal Momento: Fort Marion Drawings and the Significance of Books." In *Painters, Patrons, and Identity: Essays in Native American Art to Honor J.J. Brody,* edited by Joyce M. Szabo, 49–70. Albuquerque: University of New Mexico Press.

———. 2007. *Art from Fort Marion: The Silberman Collection.* Norman: University of Oklahoma Press.

———. 2011. *Imprisoned Art, Complex Patronage: Plains Drawings by Howling Wolf and Zotom at the Autry National Center.* Santa Fe, NM: School for Advanced Research Press.

———. 2012. "Reconstructing History from a Fragmented Past." In *Ledger Narratives: The Plains Indian Drawings in the Mark Lansburgh Collection at Dartmouth College,* edited by Colin G. Calloway, 233–52. Norman: University of Oklahoma Press.

Viola, Herman J. 1998. *Warrior Artists: Historic Cheyenne and Kiowa Indian Ledger Art Drawn by Making Medicine and Zotom.* Washington, DC: National Geographic Society.

Weist, Katherine. 1983. "Beasts of Burden and Menial Slaves: Nineteenth Century Observations of Northern Plains Indian Women." In *The Hidden Half: Studies of Plains Indian Women,* edited by Patricia Albers and Beatrice Medicine, 29–52. New York: University Press of America.

Wong, Hertha D. 1989. "Pictographs as Autobiography: Plains Indian Sketchbooks of the Late Nineteenth and Early Twentieth Centuries." *American Literary History* 1: 295–316.

MICHAEL PAUL JORDAN is Assistant Professor of Ethnology in the Department of Sociology, Anthropology, and Social Work at Texas Tech University.

4   Life-Story Objects: Folk Art and Aging in Indiana

GERONTOLOGIST ROBERT BUTLER'S 1963 article "The Life Review" challenged prevailing notions about the role of reminiscence in the lives of seniors and argued that life review was a "naturally occurring, universal mental process" (Butler 1963, 65–76). He proposed that this reflective practice among older adults was potentially productive and allowed them to recall, reorganize, and revise their understanding of life experiences. He noted that elders, with an explanatory examination of the past, forge new meaning out of old memories, sometimes as a way to understand and perhaps come to terms with their personal history. Some seniors engage in their reconsideration of the past through the use of "life-story objects," the material items that elders keep, display, and arrange that relate to personal memories and stories. A specific subset of life-story objects go beyond persistent personal possessions and are works specifically crafted to structure and tell personal narratives.[1]

Barbara Kirshenblatt-Gimblett noted that these kinds of creations, which she called "memory objects," were a way for seniors to "materialize internal images, and through them, to recapture earlier experiences." As she observes, souvenirs are often collected because of their "future ability to call back memories"; however, memory objects "are produced retrospectively, long after the events they depict transpired" (Kirshenblatt-Gimblett 1985, 331).[2] Seniors may use these time-shifting works to elicit interest, explicate personal narratives, and share their beliefs and values. Whether arranging important photographs, painting pictures of past events, or wood-burning important names and places onto a walking stick, life-story objects often anticipate social interactions and storytelling events, which is just one aspect of their creative utility and complex role in the lives of older adults.

To reduce memory paintings, story quilts, and other forms of life-story objects to mere works of art or storytelling props fails to appreciate the complex and diverse narrative nature of these creations and the process that brought them into existence. The making and use of these objects serves multiple functions in the lives of older adults: they are objects to reflect upon; tools for explaining events and their meanings; the product of a pastime that filled the lonely hours; mnemonic devices to remind the forgetful; a meditative practice that helps seniors make sense of the past; and a material legacy to leave to family and friends. In this essay, I examine the work of two elders who create pictorial narrative scenes of recalled memories, as a form of material life review.

Bob Taylor, Memory Carver

In the fall of 2013, the Columbus Area Arts Council invited me to present a workshop about these special objects and their life-review function at the Mill Race Center, an active senior facility. While going over my presentation at the center before others arrived, a tall thin man poked his head into the room (Fig. 4.1). I invited him in and

FIGURE 4.1.
Bob Taylor carving in his home in Columbus, Indiana, 2013.

introduced myself. "What do you have there?" I asked pointing to the two boards wrapped in towels that he had brought for the show-and-tell portion of the evening's program. "I do memory carvings," he explained as he unpacked the boards. On the surface of each, he had carved detailed scenes from his childhood; the pair of boards visually told the story of a train trip in 1941 that the elder, Bob Taylor, took with his parents to Coney Island in Cincinnati, Ohio. I was impressed at the artist's mastery of his craft and the narrative details invested in these panoramic panels.

"How long have you carved?"

"Since I was eight years old or so, but didn't start doing the memory carvings until after I retired."

My head raced with questions, but the room was beginning to fill with more workshop participants. "I want to talk to you more," I said, "but I have to finish getting ready for the program. Can we talk afterward?"

Several seniors brought examples of their needlepoint, quilts, woodcarvings, and other handmade works, which they took turns sharing throughout the evening. When it came time for Bob to show his work, he stood and addressed the room, "Since I have been retired, I've been doing memory carvings of my childhood." Then he lifted the boards and placed them on two tabletop stands, revealing the carved scenes. The audience's appreciation was audible as they exhaled oohs and ahhs (Fig. 4.2). He then began to tell the story of his childhood trip to Coney Island and his process of researching and making the panels to commemorate it.

The pair of panels tells the two phases of the Coney Island trip. The first shows his family ready to board the train behind the Cummins Building in Columbus. His grandfather, brother, mother, and father are waiting with him while a brass band plays (Fig. 4.3). The second panel shows them arriving at Coney Island and includes the steamboat, the *Island Queen*, that ferried them across the river; the Ferris wheel, a roller coaster, and a merry-go-round in the amusement park are shown in the distance (Fig. 4.4). He even includes a coal barge that his father just happened to snap a picture of that day. The artist's scenes are beautiful compilations of remembered moments and discovered images.

As Mary Hufford and other scholars have noted, not all autobiographies "take the form of books," but rather some of the folk art

FIGURE 4.2.
Bob Taylor at the Mill Race Center in Columbus, Indiana, 2013.

FIGURE 4.3.
Bob Taylor's carving of the train ride to Coney Island.

FIGURE 4.4.
Bob Taylor's carving of the boat trip to Coney Island.

projects of seniors are "a kind of three-dimensional reminiscence for
their makers whereby the past bursts into tangible being" (Hufford
1984, 33). These works are not just "reminiscence" or the recalling of
the past but rather thoughtfully constructed material life stories. In
many ways, Bob Taylor's art is the product of a lifetime of carving and
remembering. He blends his gift of storytelling with his love of woodcarving to produce striking narrative scenes about special moments
when he was young.

Bob invests months into researching and designing each memory
carving, before he ever puts chisel to wood. He enjoys researching the
forgotten facts and visual elements that he incorporates into his
scenes. An important part of his creative process, the time spent researching locations, people, and events not only inform his designs
but also provide him with additional stories to tell about his panels. As
the narrator of his completed carvings, he easily shifts from sharing a
childhood memory to talking about uncovering additional information for his scenes (Fig. 4.5). In Bob's memory carvings, he unites the
disparate shards of memories and discovered details into a cohesive
recalled world, through which he constructs his tight narrative scenes
and tells his personal stories. Seniors who make life-story objects often

FIGURE 4.5.
Bob Taylor's carving of the Mission Festival at White Creek.

engage in a process of personal discovery and creative expression, which is frequently followed by a series of presentations or narrations of their creations.

Exhibiting his carvings and sharing his stories at gatherings like the ones at the Mill Race Center are important to Bob, because they provide opportunities for discussions about his life-story art (Fig. 4.6).

FIGURE 4.6.
Bob Taylor and friends at the Woodcarving Show in Franklin, Indiana, 2013.

Just as he enjoys the period of reflection and research that precedes his lengthy creative process, he relishes the occasions, when he gets to present his carvings and talk about his personal memories and research discoveries. His process of reflection, creation, and narration is a dynamic system of life review. While carving is a solitary act, the elder's work researching his memories and then later presenting his carvings are social activities. During his research or discovery phase, he goes through a period of sharing his memories and learning from others about theirs. Through carving his scenes, however, his personal recollections blend with his research findings to produce a master scene that visually references important narrative aspects (Fig. 4.7). Finally, by narrating his completed panels, his woodcarving and storytelling talents combine into a unified performance.

Kirshenblatt-Gimblett observed a similar narrative dependency between her father's storytelling paintings and his verbal narratives. She writes, "The paintings are not illustrations and the stories are not captions. They are not versions of one another. Rather, different parts of the story are told in different ways in different media to form a whole that is greater than could be achieved in words or

FIGURE 4.7.
Bob Taylor's carving of a dead whale on a train.

images alone" (Kirshenblatt and Kirshenblatt-Gimblett 2007, 381–82). Comparably, the narration that accompanies Bob's art is not due to any artistic shortcomings of his carvings but rather the emergent and dialogical nature of his art. His panels not only depict the stories that he tells but beg for narration (381).

Some folklorist might dismiss the work of Bob Taylor, because his art may seem idiosyncratic rather than traditional.[3] While his carvings are distinct and masterful, life-review art is far from unique. From Grandma Moses to Mario Sanchez, elders have created recalled scenes that relate to their personal memories and stories. Borrowing Dan Ben-Amos's definition of folklore, I approach these creations as a material form of "artistic communication in small groups" (Ben Amos 1972, 9; Glassie 1989, 88). Like Bob Taylor, Marian Sykes makes life-story objects that help her construct and convey her personal experience narratives, but rather than carving memories, she hooks rugs that are distillations of family stories (Fig. 4.8).

Marian Sykes, Rug Hooker

Alone in the front room of her home near Chesterton, Indiana, a woman works at her small rug-hooking frame. Pulling colorful woolen strips through the weave of a piece of linen fabric, the artist fills in an animated scene recalled from a greased pig contest she attended long ago at a county fair. Having sketched the event onto the fabric base of what will become a hooked rug, she plies her talents to make the scene come alive. She has already hooked the out buildings and farm animals into the rug, and today she focuses her attention on the section where the children chase the pig (Fig. 4.9). She hooks for a while and then studies her progress. Something is not quite right, so she quickly pulls the loops out and begins again. Work and rework, this is the senior's daily rug-making practice. The hours pass into days, the days into weeks, until she finishes her memory project and begins to contemplate her next rug. In making her storytelling creations, the elder edits her memories and her art to produce rugs that are both beautiful to see and pleasant to recall. This is the way Marian fills her long winter days in rural northern Indiana, making rugs and meditating on her life and family.

Like Bob Taylor's research for his memory carvings, Marian's rug making requires her to work through her memories and often-told

FIGURE 4.8.
Portrait of Marian Sykes in her home in Chesterton, Indiana, 2012.

stories (Fig. 4.10). Since every piece is the culmination of months of reminiscing, sketching, and hooking, her creative process merges remembered stories with their visual representations, which produces a unified narrative. While these pictorial texts reduce complex story lines and themes down to their essential narrative components, the maker augments her rugs through narration, providing viewers with additional facts and descriptive digressions (Fig. 4.11).

In the same way that she repurposes wool in her rugs, Marian reworks her memories into art and constructs her life stories. Just

FIGURE 4.9.
Marian's hooked rug of a greased pig contest that she attended.

as an old sweater must be unraveled and washed before it can be used
in a rug, a story must be taken apart, broken into its basic narrative
pieces, purged of nonessential or unpleasant elements, and then
hooked together into an evocative scene. Marian's projects, therefore, are not just about creating art but moreover about reworking
and making sense of the past.

Current memory research reveals that the practice of retrieving autobiographical memories is a type of problem solving through which seniors
select, change, and reorganize information to meet the specific needs of
their current situations (Staudinger 2001, 150). Ursula Staudinger points
out that life review and reminiscence are two interrelated social-cognitive
activities. Reminiscence denotes a reconstruction of "life events from
memory," and life review refers to the process of recalling personal experiences as well as evaluating and interpreting those memories (Staudinger
2001, 149–50). Marian's designing and making of a rug might start with
reminiscing about the past; however, by creating sketches and through

FIGURE 4.10.
Marian's hooked rug of Little Italy in Chicago.

the long process of hooking her scenes, as well as the subsequent narrating of her rugs, the maker has developed a powerful and effective life-review practice (Fig. 4.12).

On the surface, Bob and Marian seem to have little in common, other than that their art is inspired by memories from years ago. Nevertheless, there are other traits that these two artists have in common with many of the elders with whom I work, traits that I believe deserve closer attention. First, for most elders who make life-story art, it seems to be a solitary act—a practice that can fill the lonely times in a senior's life. Second, Bob and Marian, like many older adults, started their life-review projects in their retirement years or when they experienced some other rupture in their daily lives, such as when Marian moved to northern Indiana, where she became more

FIGURE 4.11.
Marian's hooked rug of a festival on Dickens Street in Chicago.

removed from her family. Third, both of these older adults devote a considerable amount of time to designing and making their work. Spending days, weeks, and even months working on these personal objects provides Bob and Marian with special tasks to which they can devote themselves.

Finally, these adaptive pursuits also help elders to maintain and forge personal relationships. Through exhibiting life-story objects, artists engage in a special kind of communicative conduct. The artists presented use their works to connect with others in a variety of ways, such as taking their art to a woodcarving show or local rug-hooking club; or displaying their work in their homes to serve as a visible anchor for their favorite stories. Many older adults create paintings, carvings, rugs, canes, and displays for narrative encounters (Figs. 4.13, 4.14). Using their creations, older adults connect with others as storytellers, relating meaningful experiences through words and art. Through

FIGURE 4.12.
Marian's hooked rug of a Fourth of July Celebration.

FIGURE 4.13.
Gustav Potthoff's painting of *Freedom*.

FIGURE 4.14.
John Schoolman with his colorful canes in his home, North Webster, Indiana, 2008.

these items, they can summon whole worlds, explaining how things used to be, describing relationships with friends and families, and relating details about special events and places. The union of words and objects in performance links elders to their listeners, fulfilling an overt function of these curious creations.

Marian's rugs and Bob's carvings are examples of the creative ways that elders find new purpose in their lives as creators and storytellers. Their work reveals the ways that some older adults strive for satisfaction and purpose in their lives through art and story. While using the arts in elder care has become common over the past thirty years and clinical therapies that encourage reminiscence are now routine, further research into vernacular forms of material life review is needed, and folklorists have much to contribute to this creative aging movement.[4] In the 1980s and 1990s, folklorists produced several exhibits, documentaries, and academic studies that looked at the folklore of older adults, but today few in folklore focus on this topic.[5] I contend, as Simon Bronner recently observed, a "concerted discourse on the cultural traditions, practices, and beliefs of aging" is "overdue" (Bronner 2015, 7). It is my hope that we will revisit this important scholarship and revive our research interest in the memory projects of older adults and probe how these creative works help seniors in their aging process.[6]

Notes

1. This short essay previews two of the case studies and some of ideas presented in my forthcoming book in Indiana University Press's Material Vernaculars series, *Folk Art and Aging: Life-Story Objects and Their Makers*. It offers a model for studying this widespread practice and includes examples of older adults who deploy their memory projects to deal with the difficulties of aging, such as isolation, loneliness, boredom, and uselessness.

2. Barbara Kirshenblatt-Gimblett continued her research on "memory objects" of elders when she collaborated with her father Mayer Kirshenblatt on his *They Called Me Mayer July: Painted Memories of a Jewish Childhood in Poland before the Holocaust* (2007). This text marries the senior's images and stories about the Jewish community in Apt, Poland with a thoughtful analysis by his daughter, resulting in a very personal and insightful study into life-story art.

3. Federal and state arts agencies have often excluded memory art from their purview, believing that it did not conform to their tradition-based definition of folk art. For example, Robert Teske argued that "such paintings rarely constitute a traditional expressive form in folk communities and often reflect more of an individual than a communal aesthetic. Naïve paintings should therefore be omitted from future state folk art exhibitions in order to achieve greater definitional consistency and in order to avoid renewed confusion with the art historical perspective" (Teske 1985, 116).

As the above quote demonstrates, in the 1980s and 1990s, when definitional debates loomed large, many folklorists believed that memory paintings and other forms of life-story objects should not be considered "folk art." By focusing on the temporal aspect of folk art—tradition—these scholars neglected the socially and contextually situated aspects of these forms of expressive culture. John Vlach was especially vocal about not calling memory painters "folk artists," because their work lacked "clear traditional precedents and support of longstanding custom in their various communities" (Vlach 1988, 177).

4. North Dakota state folklorist Troyd Geist and a team of arts and senior-care professionals have developed an arts-in-elder-care initiative called the Arts for Life Program, which brings quilters, potters, painters, and musicians into senior care facilities. Their program has become a model arts-in-aging program. For more about this successful program, see *Art for Life: The Therapeutic Power and Promise of the Arts* (Geist 2003).

In addition, David Shuldiner once worked with the Connecticut Humanities Council as a scholar in residence. He developed a program using his skills as a folklorist to facilitate educational programs for seniors. For more information about the Connecticut program, see David Shuldiner's "Promoting Self-Worth among the Elderly" in *Putting Folklore to Use* (Shuldiner 1994, 214–25).

5. Scholars working in the 1980s brought the study of folklore and aging, as well as material forms of life review into academic discourse (Beck 1988; Bronner 1996 [1985]; Bustin and Kane 1982; Erhard 1983; Ferris 1982; Hufford 1984; Hufford, Hunt, and Zeitlin. 1987; Jabbour 1982; Kirshenblatt-Gimblett 1985; Mullen 1992; Myerhoff 1980; Shuldiner 1997), but in recent years fewer scholars have published on this front.

6. Folklorists have long been interested in the stories, reflections, and traditional knowledge of elders but rarely raise the relationship between the recalling of memories, narrative performance, and aging. Three noteworthy works that do are *Remembered Lives* (1992), a posthumous collection of Barbara Myerhoff's writings; Patrick B. Mullen's *Listening to Old Voices* (1992); and Simon Bronner's *The Carver's Art: Crafting Meaning from Wood* (1996 [1985]). Myerhoff's collection includes some of her most influential writings on the anthropology of aging and her study of ritual and performance among Eastern European Jewish seniors. Mullen's book links reminiscence and life-review scholarship to folklore by applying Robert J. Havighurst's concept of "successful aging" and Robert Butler's positive perspective of reminiscence to his study of folklore in the lives of elders. Simon Bronner's work shows how the carving projects of old men help them cope with many of the hardships of aging. Each of these scholars recognized the importance of narrative and other forms of expressive culture in the aging process. All three scholars encourage deeper study into the expressive lives of seniors through ethnographic observation and life-story collection.

References

Beck, Jane. 1988. "Stories to Tell: The Narrative Impulse in Contemporary New England Folk Art." In *Stories to Tell*, edited by Jane Beck, 38–55. Lincoln, MA: De Cordova and Dana Museum and Park.

Ben-Amos, Dan. 1972. "Toward a Definition of Folklore in Context." In *Toward New Perspectives in Folklore*, edited by Americo Parédes and Richard Bauman, 3–19. Austin: University of Texas Press.

Bronner, Simon J. 1996 [1985]. *The Carver's Art: Crafting Meaning from Wood.* Lexington: University Press of Kentucky.

———. 2015. "Forward: Folklore for the Ages." *Midwestern Folklore* 41: 3–7.

Bustin, Dillon, and Richard Kane, dir.1982. *Water from Another Time.* 28 min. Watertown, MA: Documentary Educational Resources.

Butler, Robert N. 1963. "The Life Review: An Interpretation of Reminiscence in the Aged." *Psychiatry* 26: 65–76.

Erhard, Doris Francis. 1983. *"Everybody in my family has something from me": Older Cleveland Folk Artists (exhibition catalogue).* Cleveland, OH: Department of Aging.

Ferris, William. 1982. *Local Color: A Sense of Place in Folk Art.* New York: McGraw-Hill.

Geist, Troyd A., ed. 2003. *Art for Life: The Therapeutic Power and Promise of the Arts.* Bismark, ND: North Dakota Council on the Arts.

Glassie, Henry. 1989. *The Spirit of Folk Art: The Girard Collection at the Museum of International Folk Art.* New York: Harry N. Abrams.

Hufford, Mary. 1984. "All of Life's a Stage: The Aesthetics of Life Review." In *1984 Festival of American Folklife*, 32–35. Washington, DC: Smithsonian Institution and National Park Service.

Hufford, Mary, Marjorie Hunt, and Steven Zeitlin. 1987. *The Grand Generation: Memory, Mastery and Legacy.* Washington, DC: Smithsonian Institution and National Park Service.

Jabbour, Alan. 1982. "Some Thoughts from a Folk Cultural Perspective." In *Perspectives on Aging*, edited by Priscilla W. Johnston, 139–49. Cambridge: Ballinger Publishing Company.

Kirshenblatt, Mayer, and Barbara Kirshenblatt-Gimblett. 2007. *They Called Me Mayer July: Painted Memories of a Jewish Childhood in Poland before the Holocaust.* Berkeley: University of California Press and the Judah L. Magnes Museum.

Kirshenblatt-Gimblett, Barbara. 1985. "Objects of Memory: Material Culture as Life Review." In *Folk Groups and Folklore Genres: A Reader*, edited by Elliott Oring, 329–38. Logan: Utah State University Press.

Mullen, Patrick B. 1992. *Listening to Old Voices: Folklore, Life Stories, and the Elderly.* Urbana: University of Illinois Press.

Myerhoff, Barbara. 1980. *Number Our Days: A Triumph of Continuity and Culture among Jewish Old People in an Urban Ghetto.* New York: Touchstone.

———. 1992. *Remembered Lives: The Work of Ritual, Performance and Growing Older*, edited by Marc Kaminsky. Ann Arbor: University of Michigan Press.

Shuldiner, David P. 1994. "Promoting Self-Worth among the Elderly." In *Putting Folklore to Use*, edited by Michael O. Jones, 214–25. Lexington: University Press of Kentucky.

———. 1997. *Folklore, Culture, and Aging: A Research Guide.* Westport, CT: Greenwood Press.

Staudinger, Ursula M. 2001. "Life Reflection: A Social-Cognitive Analysis of Life Review." *Review of General Psychology* 5: 148–60.

Teske, Robert T. 1985. "State Folk Art Exhibitions: Review and Preview." In *The Conservation of Culture: Folklorists and the Public Sector*, edited by Burt Feintuch, 109–17. Lexington: University Press of Kentucky.

Vlach, John M. 1988. *Plain Painters: Making Sense of American Folk Art.* Washington, DC: Smithsonian Institution Press.

JON KAY is Professor of Practice in the Department of Folklore and Ethnomusicology at Indiana University, where he is also Director of Traditional Arts Indiana and Curator of Folklife and Cultural Heritage in the Mathers Museum of World Cultures.

5 Chiefs, Brides, and Drum Keepers: Material Culture, Ceremonial Exchange, and Osage Community Life

Introduction

This chapter discusses the bridal attire associated with *Mi-zhin*, or arranged marriage, as practiced by the Osage people in the late nineteenth and early twentieth centuries. Our combined research has identified more than one hundred photographs of traditional Osage weddings (Fig. 5.1) between the 1870s and 1930s. The abundance of visual data and the continued importance of traditional bridal attire in the contemporary Osage community were important factors in our decision to investigate the topic. What we present here is part of our larger efforts to complete a longitudinal and collaborative course of research on the topic of Osage weddings and the incorporation of wedding regalia in the Paying for the Drum ceremonies of the Ilo^{n}shka dances.[1] Central to our methodology were a series of community events to promote and facilitate the direct participation of members of the Osage Nation. These efforts brought us into close conversation with a number of organizations and individuals in the Osage Nation. They include Kathryn Red Corn, director emerita of the Osage Nation Museum and Vann Big Horse, director of the Wah-Zha-Zhi Cultural Center. Public events at each institution brought hundreds of people into communication through lectures, discussions, and workshops. A highlight was an informal community exhibition of Osage wedding attire at the Wah-Zha-Zhi Cultural Center in May 2015 (Fig. 5.2).

Our research on Osage marriage ceremonies and their associated material culture is greatly enhanced by the rich and varied oral tradition that survives today in both archival and community contexts. The Doris Duke Oral History Project, conducted in 1967–72 at the University of Oklahoma, is a particularly strong resource that provides numerous first-person accounts of the marriage ceremony from the perspective of the last women married in this manner and the elders

FIGURE 5.1.
Osage bride and her matron of honor travel to the site of the wedding ceremony. Annie Collum Others wedding. Fairfax, Oklahoma, ca. 1916. (Courtesy Osage Nation Museum, Pawhuska, Oklahoma. P01-1041.)

who worked to arrange these unions. This material is augmented by reports derived from a range of popular sources that preserve community comments and accounts of events associated with traditional Osage weddings. The rechartering of Osage wedding regalia into new social and temporal contexts is a recurrent theme in contemporary

FIGURE 5.2.
Osage Bridal Attire, temporary exhibition. Wah-Zha-Zhi Cultural Center, Osage Nation, Pawhuska, Oklahoma, May 2015. (Courtesy Daniel C. Swan.)

discourse in the Osage community, and we are indebted to many individuals for sharing their perspectives and family histories.

Osage Weddings

At an early point in their tribal history, the Osage people developed a complex wedding ceremony that incorporated important tenets of philosophy, definitions of incest restrictions, and systems of reciprocal ceremonial exchange. Data derived from archived oral history interviews, historic and contemporary photographs, and our ethnographic fieldwork document the changing roles of social ceremonies and material culture as important components in the ongoing process of heritage reconstruction and cultural reproduction in Osage society.

A diverse range of sources provides consistent notice and varying degrees of documentation for Osage weddings. An early account of an Osage wedding appears in Josiah Gregg's *Commerce on the Prairie*, based on his travels on the Santa Fe Trail between 1840 and 1843 (Moorehead 1954). Gregg emphasizes the arranged nature of these unions, the strict protection of the virtues of young women, and the use of horses or other goods of value in a pattern of reciprocal exchange initiated by the young man and his family (Moorehead 1954, 429). The nature of arranged marriage and the elaborate economic negotiations associated with Osage weddings continued to attract attention from scholarly and popular writers. Kate Burrell (1903, 89) noted that this system of arranged marriage was beginning to feel the strains of acculturation at the beginning of the twentieth century. She is also the first to mention the gifting of food to the family of the prospective bride as an important aspect of the marriage proposal. Frank G. Speck (1907, 168) provides an early ethnological note on Osage marriage that provides greater detail on the structure and form of the negotiation of the dowry and is the first non-Osage observer to mention a foot race as an important social component of the marriage ceremony.

Francis La Flesche, an Omaha student of Alice Fletcher, spent most of his career documenting the ritual practices of the Osage tribal religion that were largely moribund at the time of his fieldwork in the early twentieth century. It is fortunate for our purposes that one of the few ongoing cultural practices that he documented was

Osage marriage practice in both idealized and practical forms (La Flesche 1912, 127–28):

> The marriage customs of the Osage are clearly defined, well established, and are observed to this day by the full-bloods. . . . There are two forms of marriage that are recognized as legal [in terms of Osage social mores], one which takes place between a youth and a maiden and is called *Mi-zhin*, and another in which one or both parties had been married before and is called *O-mi-ha*. The marriageable age is reached shortly after puberty and those who have attained that period of life are known as *Tse-ga-non*, or newly grown. These young people, unless near relatives, are not allowed to mingle or even speak to one another. All are strictly guarded so that none can arrange their own marriage affairs and open courtship.

In describing the ceremony, La Flesche (1912, 128–29) documents reciprocal exchange that is focused on the distribution of horses and blankets to the extended kin of the bride and groom. He makes no mention of the elaborate bridal regalia discussed in this work. His lack of notice is assumed to stem from his failure to actually witness a traditional Osage wedding.

In the context of the *Mi-zhin*, a mother and father have a son who has just passed puberty and they want to secure a life mate for him. They call together all of their relations and discuss the matter. They look for a girl of marriageable age who comes from a respectable family within the Osage tribe, a family that is at least their equivalent within the community with parents who were married in the same traditional Osage way that they wish for their son. In speaking about her marriage of the *Mi-zhin* form, performed in September 1936, Myrtle Unap (1968, 6) provides succinct explanation of her experience:

> The boy and I never spoke to each other, although we lived in the same town, but that's why we had this wedding. We never spoke to each other. We listened to our parents. That is the Osage way. They bring two families that coincide with their background, to rear children in this Osage way, which, in my opinion, compares to the royal houses of Europe. That's the way they were married and we are told that when we are married in this fashion it's a royal marriage. That we have to raise blue-blooded children for the Osage people, and we were counseled in this fashion by the older and wiser men of the Osage tribe at that time.

On the girl's side, the family is passive in that they cannot seek out a male for their daughter's mate. They must wait to be approached by some boy's family (La Flesche 1912, 127). If a proposal were accepted for consideration, the next step would be for the groom's family to move close to the home of the bride's parents. A substantial camp that often included a large circus tent and several smaller shades and shelters was established in walking distance from the home of the bride or a close relative.

From this camp, the groom's family and relations would cook meals three times a day and send this food to the home of the bride's family (Fig. 5.3). Traditionally, this went on for four days and was designed to show the bride's family what they could expect the groom to provide for their daughter throughout their life together. On the fourth day of sending food, horses were taken over to the bride's family. When the horses are accepted and distributed to the relations of the bride, this signals that the proposal has been accepted and preparations were made for the wedding ceremony to take place on the following day (Speck 1907, 167–68; Unap 1968, 1–2; Waller 1978, 10; Whitehorn 1969, 13).

FIGURE 5.3.
Members of the groom's family take food to the home of the prospective bride. Mae Vest wedding, Pawhuska Indian Village, ca. 1938. The temporary camp of the groom's family can be seen in the background. (Courtesy Jim Cooley.)

The completion of final negotiations between the two families quickly led to preparations for the wedding ceremony. Like the groom's family, the bride's family has been preparing for this event by gathering together bridal outfits (Figs. 5.4, 5.5) worn to the site of the wedding and presented to the groom's female relatives (Harper 1968, 11–12;

FIGURE 5.4.
Osage bridesmaids. (L–R) Maggie Morrell and Marian Coshehe, ca. 1930.
(Courtesy George Weston.)

FIGURE 5.5.
Osage wedding hat and coat, ca. 1930. (Courtesy Native Arts Department, Denver Art Museum. 1963.169, 157, 171 a-b.)

Hill 1967, 10; Unap 1968, 3; Whitehorn 1969, 16–17). These outfits consist of swallowtail military-style coats decorated with embroidery and epaulets and secured with wide finger-woven belts of yarn and beads. Osage brides also wear top hats surrounded with multicolored hat plumes held in place by German silver bands and ribbons. The brims of hats were often covered with brightly colored hackle feathers

(Hill 1968, 7–8; Unap 1968, 2–3). Underneath this outerwear, brides wore a set of traditional Osage women's attire comprising skirts and half leggings of wool trade cloth, silk blouses, finger-woven sashes, and German silver, pierced brooches. Not only are these bridal outfits very elaborate and costly, but the bride and bridesmaids may be wearing two or three layers of this clothing to give away to the groom's female relatives.

On the day of the wedding, the bride (Fig. 5.6) is brought down and dressed by a female relative, usually her paternal aunt, who is also the matron of honor. After the dressing of the bride and her bridesmaids, a procession is formed and the bride is placed on a prize horse with another prize horse led beside her. These horses are covered with ribbon-work blankets, beaded martingales, and German silver headstalls. The horses were eventually replaced by a carriage (Fig. 5.7) and then by automobiles. The bride is lifted off the horse and placed on a blanket and carried into a covered arbor where the wedding dinner will take place. The female relatives of the groom then undress the bridesmaids. After all are seated for the dinner, the town crier calls out for the groom to come out, and

FIGURE 5.6.
Annie Collum Others wedding, Fairfax, Oklahoma, ca. 1916. The bridesmaids and other female relatives escort the bride to a waiting wagon. (Courtesy Osage Nation Museum, Pawhuska, Oklahoma. P01-1043.)

FIGURE 5.7.
The bride and her matron of honor, wrapped in elaborate ribbon blankets, are
seated in the carriage. The blankets will be given to female members of the
groom's family. Annie Collum Others wedding. Fairfax, Oklahoma. ca. 1916.
(Courtesy Osage Nation Museum, Pawhuska, Oklahoma. P01-1040.)

he is seated next to his bride. When they eat together, the marriage is
completed, and the newlyweds are then counseled by tribal elders.
They are told what is expected of them as a traditional Osage couple,
and why they were married in this manner. The bride and groom
then spend their wedding night at the home of the groom's parents
(Hill 1968, 1–2; Unap 1968, 2–3; Whitehorn 1969, 15–17).

Chiefs Coats

In 1804, President Thomas Jefferson introduced a new material
element in US government relations with Native American nations
and their leaders. Prior to the initiation of the Lewis and Clark
Expedition, the President invited delegations from the known tribes
of the lower Missouri River Valley to Washington. Among the first
Indian Nations from the Plains region to visit the White House was
a delegation of Osage led by Chief Pawhuska of the Great Osage
(Moore and Haynes 2003 135; Viola 1995, 172). Research on the
material culture associated with the Lewis and Clark Expedition
(Moore and Haynes 2003, 135) indicates the members of the visiting

Osage delegation were outfitted with military-style clothing in St. Louis prior to the train ride to Washington, DC.

French artist Charles Balthazar Julien Févret de Saint-Mémin sketched the portraits of Pawhuska and another member of the Osage delegation and accurately captured important details of a field artillery officer's coat brought into service in 1803 (Moore and Haynes 2003, 135; Viola 1995, 172).[2] Native delegations visiting Washington, DC, in the early nineteenth century received a range of gifts that included "chiefs" coats, hats, cotton shirts, silk handkerchiefs, gloves, hair combs, clothes brushes, sashes, and other items of tailored clothing (Viola 1995, 117–18). These were important material symbols of the assimilationist policies that dominated federal relations with Native Nations and their members. These were often the final gifts the delegations received when they reached St. Louis on their journeys home (Viola 1995, 117–18). The Osage were not unique in their desire for manufactured clothing and military uniforms specifically, nor were they the only community to modify and augment these garments to reflect individual and group identities as documented among indigenous groups in the northeast in the seventeenth and eighteenth centuries (Becker 2005, 762–64; 2010, 160, 175–76).

This use of military officers coats is described by Herman Viola (1995, 120):

> Instead of the so-called citizens clothing, which became standard issue after the Civil War, the Chiefs much preferred military uniforms. Giving uniforms to Indian leaders was, like the distribution of peace medals, a carryover from the colonial era. Although the practice was at variance with efforts to civilize and Christianize, the delegates with their warrior traditions so prized the uniforms and military trappings that pragmatists in government continued the custom well into the nineteenth century. When distributing military insignia and equipment, government officers generally made some attempt to distinguish the chiefs according to their tribal status. For instance, the coats of full chiefs would carry two epaulets, half chiefs one, and warriors none.

The presentation of military coats, tailored clothing, and top hats became standard practice for Native American delegations to Washington, DC.[3] These gifts are well documented in numerous historic paintings and photographs (Figs. 5.8, 5.9). The use of military coats and hats by the Osage and members of other tribal communities is also documented in Native drawings from the nineteenth century (Fig. 5.10). Osage oral tradition consistently states that the coats presented to the visiting chiefs were far too diminutive for the stature of Osage men in the early nineteenth century and that the coats were immediately given to their daughters as symbols of rank and prestige.[4]

FIGURE 5.8.
William S. Soule (1836–1908). "Bird Chief, Arapaho," ca. 1870s. Bird Chief wears a Hardee hat, standard issue for US infantry beginning in 1858 and continuing through the Civil War. The bugle ornament was an infantry insignia. In military issue, the hat was pinned up on one side and was completed with a long ostrich plume. (Courtesy National Anthropological Archives, Smithsonian Institution, 01622802.)

FIGURE 5.9.
Cree men and traders at Fort Pitt, Saskatchewan, 1884. Trading furs for top hats, plumes, ribbons, and blankets. Osage wedding hats are an elaboration of this decorative treatment. (Courtesy Oceanco Ltd., Boca Raton, Florida.)

FIGURE 5.10.
Bad Eye (Bird Chief), Kiowa. "Osage and Kiowa," ca. 1877. (Taken from a book of drawings produced by Bad Eye during his imprisonment at Fort Marion, Florida. Fairbanks Museum and Planetarium. 7743-plate 45.)

A range of stories and interpretations continue to circulate in the Osage community regarding the manner in which military-style coats were introduced among the Osage people. Common themes include a range of colonial encounters, often attributed to the French and direct gifting by Lewis and Clark. An alternate interpretation is provided by Marguerite Waller (1976, 2B):

> Wah-Ti-Anka and Edward Chouteau left on horse back on the slow trip that was to take one and half years. Wah-Ti-Anka had a daughter who was nearing marriage age and would attain this during his absence. The pair was called to attend an event on a foreign embassy ship. During the meeting it is said that the Osage observed a military coat worn by an officer he thought to be the ship captain. As the men made ready their departure from the ship, a gesture of appreciation was extended to the Osage. Embassy officials asked if they could present Wah-Ti-Anka with something to show their appreciation. He remembered the coat and pointed to it. The officer is said to have willingly removed his coat and hat and handed them to Wah-Ti-Anka. As the months passed the men made their way back to Osage country. They stopped and washed up just outside of camp. Wah-Ti-Anka donned the coat and the plumed hat that went with it and the men rode into camp. It was his prized possession. Shortly after his arrival, Wah-Ti-Anka's daughter was to be married. Before the wedding it is said he explained that the coat meant more to him than anything he possessed and to show his love, he gave it to his daughter as part of her dowry. The daughter wore the coat and hat as her wedding apparel. Since that time, the military coat and the plumed hat has been the traditional Osage wedding apparel.

This story is interesting on a number of points. *Wahtianka* is a major culture hero in the Osage community and renditions of his travel to Washington, DC, to investigate the Americans by visiting the nation's capital are prominent. Among his many celebrated deeds is the belief that *Wahtianka* was responsible for the selection of the Oklahoma reservation and the multiple bounties that it has provided.[5] It is important to note that in 1806 members of an Osage delegation did visit the frigate USS *Adams* accompanied by President Jefferson and the Secretaries of the Army and Navy (Viola 1995, 137). Despite variances in the individuals and circumstances, the historical narrative regarding the adoption of military coats and plumed hats as Osage bridal attire likely developed from multiple origins and material transfers.

The gifting of tailored clothing, particularly military-style coats, by the US government continued into the late nineteenth century. In November 1873, Osage agent Isaac T. Gibson forwarded a letter to the Commissioner of Indian Affairs to accompany "a schedule of seventeen Osage Indians together with their measurements for coats to enable the office to fulfill a promise made by the Commissioner on his late visit to Indian Territory" (Commissioner of Indian Affairs 1873, 00659). These were likely "matchcoats," a style of clothing that evolved simultaneously with the development of military uniforms during the early colonial period (Becker 2010, 163, 165, 175). It is assumed that these were military-style coats and that the seventeen "Osage Indians" were likely the headsmen of Osage residence bands in the early 1870s (Commissioner of Indian Affairs 1872, 246).

Osage Bridal Attire

One aspect of Osage wedding attire that has escaped serious discussion is the process through which the coats themselves have been obtained over the history of their use in the Osage community. We suggest that wedding coats and hats are a previously ignored genre of Osage folk arts. While external sources have long provided coats used in Osage wedding attire, the majority would appear to have been created locally by members of the Osage community. After reviewing more than one hundred original photographs of Osage weddings (Fig. 5.11) between about 1870 and about 1935, we have never verified the use of an actual US military uniform piece as an Osage wedding coat.[6]

In the twentieth century, numerous methods were employed to obtain wedding coats long after the early nineteenth century "chief's coats" were first made available. Rose Albert Hill (1968, 7) explains one method for obtaining Osage wedding coats in her comments, "I had one made and I had it made in Wichita. But I had to send another one with it for a pattern, you know." The Fruhauf Uniform Company in Wichita, Kansas, has been identified by a number of current community members as a source of wedding coats (Shackleford 2014; Shaw 2015). Another source was (Fig. 5.12) the M. C. Lilley Company of Columbus, Ohio, a national supplier of uniform garments and accessories for military and civilian markets from 1856 to 1953 (Autry Museum 2015). The 1891 Lilly catalog included several styles

FIGURE 5.11.
Osage bridesmaids, about 1920. (Courtesy Jim Cooley.)

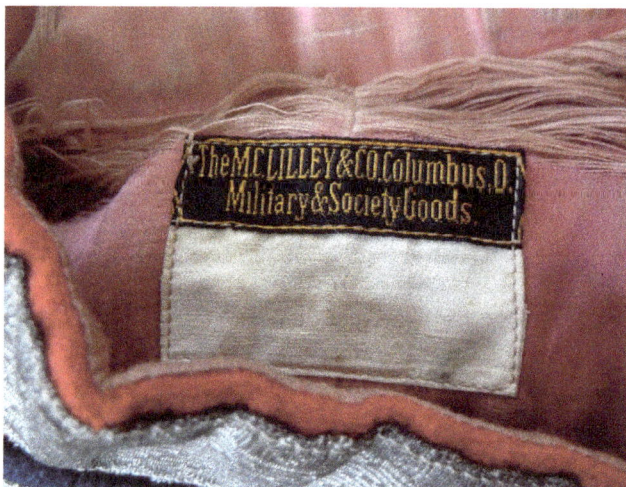

FIGURE 5.12.
M. C. Lilley Company label. Inside collar, Osage Wedding Coat, ca. 1930. The
M. C. Lilley Company was a major provider of uniforms for military and civilian
markets from 1865 to 1953. (Courtesy Renae Brumley.)

of epaulets, buttons, braid, chevrons, hat plumes, and military uniform coats (US Militaria Forum 2008, 1–3).

Marguerite Waller introduces another manner in her description of the activities of the bride's family following the acceptance of the gift of horses from the groom's family in the 1930s: "Now real plans are begun. Which takes three to four weeks. Women of the girl's family start sewing wedding clothes. There is a beehive of activities in both homes for her family and his family" (Waller 1976, 10). Her comments are indicative of a general pattern that emerged in the twentieth century that extended the date from the acceptance of the proposal to that of the actual wedding, from four days to several weeks.

In contrast to the multiple manners in which families prepared for the reciprocal exchanges associated with Osage weddings, the period to prepare for the transfer of the drum in the *Iloⁿshka* society is more clearly defined. The drum is passed during the final session of the dance but not paid for until the first session of the dance the next year. Kugee Supernaw (2015) recounts the process undertaken by his extended family to prepare for his nephew Billy Proctor to pay for the drum in the Hominy District in 2009:

> My sister, she gave five coats. And I do know that she gave away fourteen yarn belts. Six sets of five brooches. Three for the blouse and two for the coats, you know they put brooches on the shoulders. And she gave away forty Spanish embroidered shawls. We started on those coats over a year ahead of time. She [my mother] made them from scratch. There are so many tricks to that that she showed us. All kinds of stuff that I didn't even know that you had to do to build a coat. She was a seamstress; she worked at it all her life. She started as a teenager working at Miss Jacksons in Tulsa and then she had her own dress shops, several of them over the years. But she did it all her life and she was really good at it. These were her first wedding coats. We went down and got some Simplicity patterns for winter coats and she just altered them. She came up with all of the different designs. I don't remember her using pictures for inspiration. They all came from her mind. We made the hats. Billy did that. I bought those stove pipe hats on the Internet. I made the silver bands and [on] some we may have used commercial, stamped brass. Billy did most of the feather work. He tied them, one at a time.

The bridal outfits created by the Supernaw family (Figs. 5.13, 5.14) provide excellent examples of modern interpretations of Osage

FIGURE 5.13.
Osage "brides," Paying for the Drum ceremony, Hominy District, 2005. Each bride wears a unique hat and coat designed and produced by various members of the Supernaw family. The brides also wear a new set of traditional Osage woman's attire under the coats. (Courtesy Billy Proctor.)

FIGURE 5.14.
Osage "brides" (reverse), Paying for the Drum ceremony, Hominy District, 2005. The backs of the wedding coats are decorated with elements drawn from the traditional dress of Osage women. This includes sets of German silver brooches decorated in a pierced technique and ribbon drops in various configurations. Fine finger-woven sashes with traditional Osage designs are used to secure the wedding coat. (Courtesy Billy Proctor.)

wedding attire. They reflect the influence and contribution of the designer and tailor of the coats, Irene Supernaw, the grandmother of Drum Keeper Billy Proctor. We are astounded by the creativity and skill reflected in Osage wedding coats constructed in the twentieth and twenty-first centuries. It is common for people to use commercial patterns as the foundation for the construction of coats with modifications and embellishments, including novel approaches and designs (Fig. 5.15). The Wah-Zha-Zhi Cultural Center sponsors periodic workshops on the construction of hat plumes and wedding hats (Fig. 5.16) and is preparing to offer a session on wedding coats.[7]

FIGURE 5.15.
Osage wedding coat (detail), ca. 2000. A modern example with epaulets fashioned from oak tan leather, sash cord, and Chainette fringe. (Courtesy Sam Noble Museum, University of Oklahoma. E 2014.1.10.)

FIGURE 5.16.
Osage wedding hat. Given to Robert Harris by Bates Shaw when he paid Robert for the drum in the Gray Horse District, ca. 2000. (Courtesy Renee Harris.)

The *Ilo^{n}shka* Society and Paying for the Drum

In the early 1880s, the Osage were exposed to a number of revitalization movements including Peyotism, the Ghost Dance, and Christianity (Granberry 1987; Swan 1990). One of the most important of these is the Osage *Ilo^{n}shka* (Fig. 5.17). The *Ilo^{n}shka* is the Osage version of the Grass Dance, a major American Indian cultural movement adopted by dozens of tribes in the Plains and Prairie areas in the late nineteenth century (Wissler 1916). While not a religion and certainly more than a social dance, the *Ilo^{n}shka* is a ceremonial society of male members that preserves and perpetuates many traditional values of the Osage while

FIGURE 5.17.
Osage *Ilo*^{n}*shka* Society, Gray Horse, Oklahoma, 1912. Polygonal, wooden frame
dance houses and large, native-constructed drums are iconic in the material culture
of the Grass Dance in Prairie and Plains communities. (Courtesy Jim Cooley.)

providing a focal point for the modern expression of individual and
group identities. The *Ilo*^{n}*shka* has been an important part of the cultural
life of Osage people for over a century. The dance and its associated
organization provide a spiritual charter for the survival of the ancient
Osage physical divisions, or "districts" as they are called today. The
Ilo^n*shka* developed among the Osage along district lines, with a separate
drum and dance organization established in each physical division of
the tribe (Lookout 1998).

As a formal society, the *Ilo*^{n}*shka* is organized around a committee
comprised of ceremonial officers and an initiated membership. At the
head of the organization is the drum keeper, charged with the care and
protection of the drum and sponsorship of the annual dance in his
respective district (Callahan 1993, 34–50). Drum keepers traditionally
hold the drum for a self-selected period, generally four or five years,
with several drum keepers in the mid-twentieth century holding it for
ten years or longer. In addition to the symbolic act of providing physical care and protection for the drum, the most important work of the
drum keeper and his committee is to sponsor, organize, and conduct

the annual dances. The four-day ceremony provides the main opportunity for the Osage districts to reconstitute their community identities
as they work to host their counterparts from the other two districts. The
transition or "passing of the drum" from one drum keeper to another
creates new sets of social relationships through the reformulation of
the *Ilo[n]shka* committees, providing important continuity while accom-
modating change with each new permutation.

The newly selected drum keeper and his extended family will
work over the course of the coming year to prepare the necessary
goods, gifts, and food that will be distributed when he pays for the
drum and sponsors his first set of dances. The Paying for the Drum
event begins when the committee and society members from the
district assemble before the start of the dances on Thursday afternoon. After the dancers from the visiting and host districts have been
seated in the arbor, the new drum keeper and his family file in and
arrange themselves before the assembled members of the *Ilo[n]shka*. A
number of trunks and bundles are brought into the arena and several
of the new drum keeper's female relatives are dressed in traditional
Osage wedding outfits (Fig. 5.18).

FIGURE 5.18.
Paying for the Drum ceremony, Pawhuska, Oklahoma, ca. 1948. Osage "brides"
(L–R) Margaret Gray, Dora Lookout, and Maggie Morrell. (Courtesy Gina Gray.)

Today it is common for six to eight "brides" to be "given away" in the Paying for the Drum event. The first individual that is recognized in the giveaway is the former drum keeper, who receives the traditional gift of a horse with a blanket draped over its back. Common recipients of a "bride" are female relatives of the former drum keeper and the current headman, cooks, and advisors. The brides are escorted out of the dance area where a number of recipient's female relatives "undress the bridesmaid," who is stripped to shorts and a T-shirt with the garments and accoutrements carefully folded and put away. Each member of the newly formed committee is recognized and receives a gift from the drum keeper, often a shawl or a Pendleton blanket (a highly valued woolen blanket produced by Pendleton Woolen Mills). The drum keeper completes the Paying for the Drum ceremony with a sizable monetary gift to the Committee Singers.

The last traditional Osage weddings took place in the mid-1930s, creating what we believe to have been an abundance of the bridal outfits in the Osage community. These outfits were highly regarded for their beauty, cost, and importance as objects of familial patrimony and physical evidence of a more "traditional" form of Osage society. At some point, a conscious decision was made to insert these wedding outfits into the $Ilo^n shka$ to broaden their role in the reproduction of Osage society. When this occurred is unknown and will likely remain so. No reference to the specifics of this innovation appears in any of the Doris Duke interviews recorded in the 1960s among the Osage elders at the time, either in their accounts of the marriage ceremony itself or of the $Ilo^n shka$ dance.

Based on three years of research, we were initially prepared to conclude that the incorporation of bridal attire into the $Ilo^n shka$ dance occurred after World War II, an important period in the history of the dance when women assumed greater participation in both the governance and performance of the dance. In looking at the parallels between the Osage wedding and the Paying for the Drum ceremonies, one could argue that this payment represents a "symbolic marriage" between the extended families and committees of the outgoing and incoming drum keepers.[8] We have recently become aware of multiple lines of evidence that indicate this incorporation occurred prior to World War II and perhaps as early as the mid-1920s, when traditional Osage weddings achieved exceptional levels of gift exchange (Julia Lookout, pers. comm. March 25, 2015; Shackleford 2014).

The opportunity to discuss the experience of paying for the drum with a number of drum keepers and their families emphasized a shared aspect of historical consciousness in which the memory of a year of fervent activity and preparation was eclipsed by the intense experience of giving away the products of that effort in less than an hour. The experiences of incoming drum keepers provide an interesting opportunity to document the respective roles of preparatory process and ritual performance in expressive culture. Preparing to pay for the drum and sponsorship of the annual *Ilo^nshka* dances provides the greatest level of familial and community integration and cooperation, critical in the construction and maintenance of social relations.

In addition to the rechartering of Osage wedding attire as an important economic and symbolic element in the Passing of the Drum ceremony in the *Ilo^nshka* society, it has also became emblematic of Osage national identity in external relations with other Native communities. It is common for the Osage Nation princess (Fig. 5.19) to wear bridal attire during the American Indian Exposition in Anadarko, Oklahoma, one of the largest intertribal events in the state.[9] Swan has also seen wedding coats used to celebrate the seventy-fifth anniversary of the Osage Nation Museum (in 2013) and in fashion shows (1995 and 2012) that featured both traditional and modern ensembles created by Osage designers. The movement of the wedding regalia into broader realms of Osage art and society is evidenced by a miniature Christmas tree (Fig. 5.20) decorated by Anita Lookout. The tree features ornaments inspired by traditional Osage arts and benefits from Mrs. Lookout's expertise as a finger-weaving artist.

Discussion

The goal of our broader research on Osage wedding attire is to produce a North American complement to the work of Andrew Strathern and Pamela Stewart (2005a, 2005b) and Annette Weiner (1980) on the intersection of economic exchange, ritual performance, and intellectual property in the reproduction of social relations in contexts of external change. The concept of circulation was key to the initial diffusion of social movements like the Drum Religion and Grass Dance

FIGURE 5.19.
Thomasine Greene, Osage tribal princess, 1949. American Indian Exposition
parade, Anadarko, Oklahoma. (Courtesy Oklahoma Historical Society.
20912.16.9.)

with success and longevity of these diffused and exchanged cultural
forms predicated on the continued circulation of drums and the goods
used to "pay" for them. The transition of the Dream Dance among
the Menominee into the modern Powwow (Slotkin 1957) is one example among many in which these movements became impossible to
sustain as the ability to introduce the dance (drums) into new locales
declined.

   The autonomy of the three Osage districts and their respective
cycles of passing the drum, the reciprocal hosting and feasting among
the *Ilo^nshka* districts, and the distribution of rations and the gift exchanges associated with the family songs in each district are key factors

FIGURE 5.20.
Miniature Christmas tree with Osage-inspired ornaments. Anita Lookout, Pawhuska, Oklahoma, 2015. Mrs. Lookout was a renowned finger-weaving artist in the Osage community. (Courtesy Sam Noble Museum, University of Oklahoma. E-2015.1.48.)

in the cultural production of contemporary Osage society. The gifting of Osage "brides" as an integral component in the perpetuation of the *Ilo^{n}shka* is an important example of the rechartering of an existing material culture to perpetuate traditional values and key intellectual property in response to changing economic and political circumstances.

Systems of ceremonial exchange have been a foundational topic of special interest in social anthropology, and they have been a focus of work in several regions of Native North America. Strathern and Stewart (2005a, 230) provide an effective general conception of ceremonial exchange, one that articulates with the details of the *Ilo^nshka* as a Native American example of this widespread form of cultural production: "Ceremonial exchange is a term that anthropologists have applied to systems in which items of value are publicly displayed and given to partners on a reciprocal basis over time. Typically, these occasions are marked by dancing and festivities, where men, women and children participate in one way or another. This involvement of the community demonstrates the social importance of the complex events involved. These events also create and maintain forms of political alliance between partners, whether these are particular persons or groups." The Osage *Ilo^nshka* is just such an event. Built around ritualized gift exchange, the *Ilo^nshka* is a festive event that includes dancing and feasting as key elements. As noted by Strathern and Stewart (2005a, 230), such gift-centered ceremonials are "an important constitutive factor in the political order of society."

Wedding coats also accrue significance and value beyond their monetary costs to produce. While the exact timing and circumstances of the incorporation of military-style coats in Osage wedding regalia in the early nineteenth century will likely remain elusive, the intent is clearly stated in the following interpretation ("Osage Wedding Coat" 1998):

> Traditional Osage Wedding coats are taken from the days of the Osage involvement in regulating the river systems of the Missouri River with the foreign powers of Spain, England, France and eventually the United States where it was common practice to give officer's jackets to the leaders of the Osage in exchange for some privilege or favor. The Osages treasured these jackets and when paying for something of great importance, or in this case a bride, gave them away to show the importance of the event. The Wedding Coat remains the preeminent form of gratitude among the Osages and when families pay for the right of keeping the drum; often offer several Wedding Coats.

Viewed in this manner, Osage wedding coats provide an excellent example of the role of material culture in the reproduction of Osage social relationships in multiple sociotemporal contexts. As explained by this anonymous source, the gifting of wedding outfits constitutes the ultimate material form of respect in a society in which status is largely governed by the tenets of hospitality and generosity.

Strathern and Stewart (2005b, 3) expand their discussion of the role of gift exchange in the reproduction of social relations to address the expressive forms central to the contextualization of social values and practices in situations of external change:

> Expressive activities, such as songs, dances, folktales, rituals, art, and literature, are often referred to as revealing the inner sensibilities of people and their forms of cultural continuity, It is clear that sensibilities undergo change and are also creatively reshaped over time. Also, continuity and change may be simultaneously expressed in a given artistic medium. While such media may be claimed by their producers and consumers as representing tradition, they may actually be eloquent witnesses to historical change. Expressive genres thus become a particularly poignant context for examining classic problems of cultural adaptation, bricolage, historical consciousness, and assertions of identity, both personal and collective, whether these are seen in local or translocal terms.

Osage wedding regalia and its ritual exchange in different contexts certainly illustrate the core concepts associated with the challenge of maintaining a sense of tradition and social stability in situations of disruption and accelerated change. In the twentieth century, the elements of change that challenged social stability among the Osage included the forced allotment of their reservation, political domination by the US Senate, and the challenges often associated with rapid economic affluence. The evolution of Osage bridal attire exemplifies the concept of "bricolage" and their circulation facilitates intersections in the life cycles of individuals and objects of exchange. Osage wedding outfits provide physical evidence of the multiple contexts in which systems of interaction contribute to the development and maintenance of social relationships through the "reproduction and regeneration of certain elements of value" (Weiner 1980, 83). Among the Osage, these elements of value include the *Ilo^{n}shka* dances, reciprocal hosting and feasting, and Osage "brides."

Osage wedding attire holds tremendous potential to contribute to a broader scholarship on cloth and clothing as material culture. Here, our work resonates with that of Chloë Colchester on the role of material objects as stabilizing elements in times of external change. Her work (Colchester 2005, 139) on the revival and repositioning of chiefly dress in Colonial Fiji documents an analogous situation in which a physical marker of chiefly status and prestige is used to validate and structure new historical contexts.

Conclusion

Osage wedding regalia combines a range of European garments and materials modified and expanded according to an indigenous aesthetic to create a unique symbol of power and prestige. The regalia have been deployed in a range of social contexts over the past two hundred years. Dozens, if not hundreds, of these coats and hats reside in personal and family collections in the Osage community (Fig. 5.21), considerably more than are present in museum collections. Community members greatly value these coats, both historic and contemporary in provenance. Vintage coats and hats are treasured objects that invoke the memory of the female ancestor who once possessed them. The stories associated with Osage wedding coats emphasize the identity of the recipients of these gifts as opposed to the identity of the individuals who gave them. As such, they continue their role as prestige objects, evidence of the status necessary to receive the special recognition of the gift of a "bride" at a traditional wedding. The biography of these objects is well preserved and includes the names of the recipient and her relationships with the families joined through these unions. These coats and hats provide material evidence and markers of the social, political, and economic relationships established through traditional Osage marriage. These vintage coats are objects of heritage that embody an idealized past. They represent early nation-to-nation intercourse as chiefs from the Osage Nation gained audiences with European and American political authorities.

Osage wedding regalia has long functioned as a unique material attribute and symbol of Osage identity and community. These outfits represent the ostentatious and extravagant nature of Osage expressive culture, both deserved and perceived. These wedding outfits have found no purpose or use outside Osage society, avoiding adoption by other communities in the cultural milieu of Oklahoma in the late nineteenth century. Osage wedding regalia, once placed within the context of the Paying for the Drum ceremony of the $Ilo^n shka$ dances, is protected from the diffusional forces of powwow culture.

FIGURE 5.21.
Danielle Cass wearing the wedding outfit gifted to her brother, former Pawhuska Drum Keeper Bruce Cass, by William Shunkamolah when he paid for the drum in the Hominy District, 2009. The sash securing the coat is unfinished, a traditional characteristic of wedding belts that is perpetuated by some drum keepers. Wah-Zha-Zhi Cultural Center, Pawhuska, Oklahoma, 2015. (Courtesy Daniel C. Swan.)

In this context, modern coats symbolize the special status of drum keepers and those who support them. Their exchange (gifting) provides public expression of bonds of friendship and respect and marks the transfer of authority and responsibilities associated with leadership positions within the *Iloⁿshka* society.

Osage people have always invested creativity and community aesthetics into the construction of wedding coats and hats. The use of military officers coats is the beginning of a long tradition of Osage wedding coats. In all cases, these garments, either commercially produced or locally constructed, were embellished with ribbons, metal brooches, and woven sashes that adhere to community preferences and standards. The Osage people have employed diverse methods to secure wedding regalia over the past years. In this process, they developed what we consider a previously unconsidered genre of American Indian art. Regardless of the methods employed and the temporal context, the assembly of multiple Osage wedding outfits is an undertaking that requires considerable labor and the resources of an extended kin network. This is evident in the historical context of Osage weddings and in the modern $Ilo^n shka$ society.

The autonomy of the three Osage districts and their respective cycles of passing the drum, the reciprocal hosting and feasting among the $Ilo^n shka$ districts, and the distribution of rations and the gift exchanges associated with the family songs in each district are key factors in the cultural production of contemporary Osage society. The passing of the drum and the restructuring of the $Ilo^n shka$ committee provide important opportunities for the dance to maintain its cultural integrity while making room for realignments and reinterpretations. The gifting of Osage "brides" is a critical factor in the perpetuation of the $Ilo^n shka$ and an important example of the rechartering of a previous existing material culture to perpetuate traditional values and key intellectual property in response to changing economic and political circumstances.

Acknowledgments

We thank the many individuals who supported and contributed to our research. We express our appreciation to Chief Geoffrey Standing Bear for his support and encouragement for the larger research exhibition and publication project. We are particularly indebted to Kathryn Red Corn, director *emerita* of the Osage Nation Museum and Vann Bighorse, director of the Wah-Zha-Zhi Cultural Center. We also wish to acknowledge the support of Hallie Winter, director of the Osage Nation Museum, and Shannon Shaw Duty, editor of the *Osage News*. We also appreciate the invaluable support and assistance from the staffs at the Osage Nation Museum and the Wah-Zha-Zhi Cultural Center. We acknowledge the many individuals who shared stories, photographs, heirlooms, and memories with us during the course of our research. Our research received financial support from the Sam Noble Museum and the College of Arts and Sciences at the University of Oklahoma. The authors alone remain responsible for the content of this work and any misrepresentations or misinterpretations presented herein.

Notes

1. For a full discussion of the Osage *Ilo^{n}shka* the reader is referred to Callahan (1993), Feder (1980), and Mathews (1961).

2. To view this image of "Payouska" by St. Memin, the reader is directed to http://osageweddings.com/osage-wedding-regalia/military-coats-and-native-americans/.

3. To view photographs and paintings of Native Americans wearing military style coats the reader is directed to http://osageweddings.com/osage-wedding-regalia/military-coats-and-native-americans/.

4. The exceptional height and physical stature of adult Osage males is well documented throughout the historic period. See Bradbury (1819, 42), Catlin (1973, 40), and McDermott (1940, 136).

5. The Osage reservation in Indian Territory, the current state of Oklahoma, has provided consistent sources of revenue for the tribe and its members, including the lease of pastureland in the late nineteenth and early twentieth centuries, and the discovery and extraction of substantial oil and gas deposits in the twentieth and twenty-first centuries. See Bailey (1970), Mathews (1961), and Wilson (1985) for in-depth treatment and analysis.

6. Based on our combined attendance at the *Ilo^{n}shka* over the past forty years we conclude that the vast majority of the wedding coats used in the Paying for the Drum ceremonies in this period were created for the specific event. Given that there are three *Ilo^{n}shka* districts, that the average tenure of drum keepers is four or five years and five or six "brides" are given in each Paying for the Drum ceremony, this creates a potential corpus of hundreds of examples for additional study. It has also become increasingly rare for a vintage coat to be included in the bridal attire gifted by an incoming drum keeper.

7. These workshops benefit from the direct experience of Cultural Center employees in paying for the drum. Director Vann Big Horse is a former drum keeper in the Pawhuska District and Program Specialist Renee Harris is the mother of former Gray Horse drum keeper, Robert Harris. The addition of a workshop on wedding coats in the winter of 2015–16 is timely with new drum keepers in the Gray Horse and Hominy districts who will pay for their respective drums in June 2016.

8. On several occasions, Swan was exposed to discussions between "Brides" and recipients in which the implied symbolism was the source of joking behavior. On one occasion Harry Red Eagle Jr. said to Lucille Roubedeaux, "Well, I guess we are married now. Don't tell my wife!"

9. To view photographs of Osage princesses in wedding regalia the reader is directed to http://osageweddings.com/osage-wedding-regalia/osage-pricesses-in-wedding-attire/.

References

Autry Museum. 2015. "The Autry's Collection Online. Saber- 93.33.2." Accessed November 28, 2015. http://collections.theautry.org/mwebcgi/mweb.exe?request=record;id=PE215964;type=701.

Bailey, Garrick A. 1970. "Changes in Osage Social Organization 1673–1969." PhD diss., University of Oregon.

Becker, Marshall Joseph. 2005. "Matchcoats: Cultural Conservatism and Change in One Aspect of Native American Clothing." *Ethnohistory* 52: 727–87.

———. 2010. "Match Coats and the Military: Mass Produced Clothing for Native Americans as Parallel Markets in the Seventeen Century." *Textile Histories* (supplement) 41: 153–81.

Bradbury, John. 1819. *Travels in the Interior of America, in the Years 1809, 1810, and 1811*, 2nd ed. London: Sherwood, Neely, and Jones. https://archive.org/details/travelsininteri00bywagoog.

Burrell, Kate. 1903. "As Osage Indians Live Today." *Sturms Oklahoma Magazine* 2(5): 84–89.

Callahan, Alice Anne. 1993. *The Osage Ceremonial Dance: I'-Lon-Schka*. Norman: University of Oklahoma Press.

Catlin, George. 1973. *Letters and Notes on the Manners, Customs and Conditions of North American Indians*. New York: Dover Publications.

Colchester, Chloë. 2005. "Relative Imagery: Patterns of Response to the Revival of Archaic Chiefly Dress in Fiji." In *Clothing as Material Culture*, edited by Susanne Kuchler and Daniel Miller, 139–58. New York: Berg Publishers.

Commissioner of Indian Affairs. 1872. *Annual Report, 1872*. Washington, DC: Government Printing Office.

Commissioner of Indian Affairs. 1873. *Letters Received by the Office of Indian Affairs, 1824–1881*. RG-75, M234. Fort Worth, TX: National Archives.

Feder, Norman. 1980. "Some Notes on the Osage War Dance." *Moccasin Tracks* 6(3): 4–7.

Granberry, Allison L. 1987. "The Expression of Osage Identity: Ethnic Unity and the In-Lon-Schka." MA thesis, University of Tulsa.

Harper, Hazel, interviewed by Katherine Red Corn. May 1968. Doris Duke Oral History Collection, Western History Collections, University of Oklahoma Libraries.

Hill, Rose Albert, interviewed by Robert L. Miller. June 5, 1967. Doris Duke Oral History Collection, Western History Collections, University of Oklahoma Libraries.

Hill, Rose Albert, interviewed by Katherine Red Corn. June 1968. Doris Duke Oral History Collection, Western History Collections, University of Oklahoma Libraries.

La Flesche, Francis. 1912. "Osage Marriage." *American Anthropologist* 14: 127–30.

Lookout, Morris. 1998. Manuscript on the Elonshka Society. Photocopy in the possession of Swan.

Mathews, John Joseph. 1961. *The Osages: Children of the Middle Waters*. Norman: University of Oklahoma Press.

McDermott, John Francis, ed. 1940. *Tixier's Travels on the Osage Prairies*. Norman: University of Oklahoma Press.

Moore, Robert J., Jr., and Michael Haynes. 2003. *Lewis and Clark, Tailor Made and Trail Worn: Army Life, Clothing and Weapons of the Corps of Discovery.* Helena, MT: Farcountry Press.

Moorehead, Max. L. ed. 1954 *Commerce of the Prairie.* Norman: University of Oklahoma Press.

"Osage Wedding Coat." May 30, 1998. *Pawhuska Capital Journal.*

Shackleford, Romaine, interviewed by Daniel C. Swan. June 25, 2014. Ethnology Department, Sam Noble Museum, University of Oklahoma.

Shaw, Jerry, and Ruth, interviewed by Daniel C. Swan. June 10, 2015. Department of Ethnology, Sam Noble Museum, University of Oklahoma.

Slotkin, J. S. 1957. Menominee Powwow. Milwaukee, WI: Milwaukee Public Museum.

Speck, Frank G. 1907. "Notes on the Ethnology the Osage Indians." *Transactions of the Department of Archaeology, Free Museum of Science and Art, University of Pennsylvania,* 2: 159–71.

Strathern, Andrew, and Pamela Stewart. 2005a. "Ceremonial Exchange." In *A Handbook of Economic Anthropology,* edited by James. G. Carrier, 230–45. Cheltenham, England: Edward Elgar Publishing.

Strathern, Andrew, and Pamela Stewart. 2005b. "Introduction." In *Expressive Genres and Historical Change,* edited by Pamela Stewart, 1–39. Burlington, VT: Ashgate.

Supernaw, Kugee, interviewed by Daniel C. Swan. June 30, 2015. Ethnology Department, Sam Noble Museum, University of Oklahoma.

Swan, Daniel C. 1990. "West Moon–East Moon: An Ethnohistory of Osage Peyotism, 1890–1930" PhD diss., University of Oklahoma.

Unap, Myrtle, interviewed by Katherine Maker. November 30, 1968. Doris Duke Oral History Collection, Western History Collections, University of Oklahoma Libraries.

US Militaria Forum. 2008. "M.C. Lilley and Company Catalogue of Military Goods 1891." Accessed. November 28, 2015. http://www.usmilitariaforum.com/forums/index.php?/topic/15954-1891-mclilley-co-catalogue/.

Viola, Herman J. 1995. *Diplomats in Buckskin.* Bluffton, SC: Rivilo.

Waller, Marguerite. 1976. "Osage Brides Keep a Tradition Alive." *Pawhuska Capital Journal* July 4.

———. 1978. "Osages Were Never Afraid of Death." In *Reflections of Early Day Hominy,* edited by David Million, 9–12. Hominy, OK: Chamber of Commerce.

Weiner, Annette. 1980. "Reproduction: A Replacement for Reciprocity." *American Ethnologist* 7: 71–85.

Whitehorn, Lillie Hoag, interviewed by B. D. Timmons. June 29, 1969. Doris Duke Oral History Collection, Western History Collections, University of Oklahoma Libraries.

Wilson, Terry. 1985. *The Underground Reservation: Osage Oil.* Lincoln: University of Nebraska Press.

Wissler, Clark. 1916. "General Discussion of Shamanistic and Dancing Societies." *Anthropological Papers of the American Museum of Natural History* 11:853-76.

DANIEL C. SWAN is Curator of Ethnology at the Sam Noble Oklahoma Museum of Natural History and Professor of Anthropology in the Department of Anthropology at the University of Oklahoma.

JIM COOLEY is Research Associate in the Department of Ethnology at the Sam Noble Oklahoma Museum of Natural History at the University of Oklahoma.

Index

While not evoked directly in its title, the concepts of material culture, making, and craft permeate the volume and are, consequently, not indexed.

www.ingramcontent.com/pod-product-compliance
Lightning Source LLC
Chambersburg PA
CBHW050809270326
41926CB00026B/4643